CRE公務員綜合招聘考試

精讀試題王 題庫與練習

投考公務員
能力傾向測試精讀王

Man Sir & Mark Sir 著

序言

現今社會的變遷和經濟的轉型為政府的施政帶來極大的挑戰。因此，公務員團隊必須吸納更多的有志者、有能者，為市民提供優質的服務。所謂有志者，簡單而言，正如2011-12年度施政報告所示：「堅守以民為本的信念，以開放包容的態度，服務市民，貢獻社會。」至於有能者則包括各方的專才，不一而足，且各部門的要求也有所不同，難以一概而論。

另一方面，專才也須具備通才的特質，據公務員事務局所示：「政務職系人員是專業的管理通才，在香港特別行政區政府擔當重要角色。」所以，公務員考試組及部分決策局和部門舉辦一系列的考試遴選，以為聘任之用。

以學位／專業程度職系而言，最基本的要求就是通過公務員綜合招聘考試（Common Recruitment Examination-CRE），該測試首先包括三張各為45分鐘的多項選擇題試卷，分別是「中文運用」、「英文運用」、和「能力傾向測試」，其目的是評核考生的中、英語文能力及推理能力。

之後是「基本法測試」試卷，基本法測試同樣是以選擇題形式作答之試卷，全卷合共15題，考生必須於20分鐘內完成。而基本法測試本身並無設定及格分數，滿分則為100分。基本法測試的成績，會對於應徵「學位或專業程度公務員職位」的人士佔其整體表現的一個適當的比重。

然而，學有博約之別，才有遲速之分，一些考生雖有志有能，但礙於此一門檻，因而未能加入公務員團隊，一展抱負。

有見及此，本書特為應考公務員綜合招聘試的考生提供試前準備，希望考生能熟習各種題型及答題方法。可是要在45分鐘之內完成全卷對大部分考生而言確有一定的難度。因此，答題的時間分配也是通過該試的關鍵之一。考生宜通過本書的模擬測試，了解自己的強弱所在，從而制訂最適合自己的考試策略。

此外，考生也應明白任何一種能力的培訓，固然不可能一蹴而就，所以宜多加推敲部分附有解說的答案，先從準確入手，再提升答題速度。考生如能善用本書，對於應付公務員綜合招聘考試有很大的幫助。

Man Sir & Mark Sir

目錄

PART
ONE

輕鬆認識 CRE

認識公務員綜合招聘考試

公務員綜合招聘考試(CRE)科目包括:

- 英文運用
- 中文運用
- 能力傾向測試
- 《基本法》知識測試

入職要求

- 應徵學位或專業程度公務員職位者，須在綜合招聘考試的英文運用及中文運用兩張試卷取得二級或一級成績，以符合有關職位的一般語文能力要求。
- 個別進行招聘的政府部門/ 職系會於招聘廣告中列明有關職位在英文運用及中文運用試卷所需的成績。
- 在英文運用及中文運用試卷取得二級成績的應徵者，會被視為已符合所有學位或專業程度職系的一般語文能力要求。
- 部分學位或專業程度公務員職位要求應徵者除具備英文運用及中文運用試卷的所需成績外，亦須在能力傾向測試中取得及格成績。

考試模式

I. 英文運用

考試模式：

全卷共40題選擇題，限時45分鐘

試題類型：

- Comprehension
- Error Identification
- Sentence Completion
- Paragraph Improvement

評分標準：

成績分為二級、一級及格或不及格，二級為最高等級

擁有以下資歷者可等同獲CRE英文運用考試的二級成績，並可豁免考試：

- 香港中學文憑考試英國語文科5級或以上成績
- 香港高級程度會考英語運用科或 General Certificate of Education(Advanced Level) (GCE ALevel) English Language 科C級或以上成績

- 在International English Language Testing System(IELTS)學術模式整體分級取得6.5或以上，並在同一次考試中各項個別分級取得不低於6的人士，在考試成績的兩年有效期內，其IELTS成績可獲接納為等同綜合招聘考試英文運用試卷的二級成績。

擁有以下資歷者可等同獲CRE英文運用考試的一級成績：

- 香港中學文憑考試英國語文科4級成績
- 香港高級程度會考英語運用科或GCE ALevel English Language科D級成績

* 備註：持有上述成績者，可因應有意投考的公務員職位的要求，決定是否需要報考英文運用試卷。

II. 中文運用

考試模式：

全卷共45題選擇題，限時45分鐘

試題類型：

- 閱讀理解
- 字詞辨識
- 句子辨析
- 詞句運用

評分標準：

成績分為二級、一級或不及格，二級為最高等級

擁有以下資歷者可等同獲CRE中文運用考試的二級成績，並可豁免考試：

- 香港中學文憑考試中國語文科5級或以上成績
- 香港高級程度會考中國語文及文化、中國語言文學或中國語文科C級或以上成績

擁有以下資歷者可等同獲CRE中文運用考試的一級成績：

- 香港中學文憑考試中國語文科4級成績
- 香港高級程度會考中國語文及文化、中國語言文學或中國語文科D級成績

* 備註：持有上述成績者，可因應有意投考的公務員職位的要求，決定是否需要報考中文運用試卷。

III. 能力傾向測試

考試模式：

全卷共35題選擇題，限時45分鐘

試題類型：

- 演繹推理
- Verbal Reasoning (English)
- Numerical Reasoning
- Data Sufficiency Test

- Interpretation of Tables and Graphs

評分標準：

成績分為及格或不及格

IV.《基本法》知識測試

考試模式：

全卷共15題選擇題，限時20分鐘

評分標準：

無及格標準，測試應徵者對《基本法》（包括所有附件及夾附的資料）的認識。成績會在整體表現中佔適當比重，但不會影響其申請公務員職位的資格。

公務員職系要求全面睇

	職系	入職職級	英文運用	中文運用	能力傾向測試
1	會計主任	二級會計主任	二級	二級	及格
2	政務主任	政務主任	二級	二級	及格
3	農業主任	助理農業主任／農業主任	一級	一級	及格
4	系統分析／程序編製主任	二級系統分析／程序編製主任	二級	二級	及格
5	建築師	助理建築師／建築師	一級	一級	及格
6	政府檔案處主任	政府檔案處助理主任	二級	二級	-
7	評税主任	助理評税主任	二級	二級	及格
8	審計師	審計師	二級	二級	及格
9	屋宇裝備工程師	助理屋宇裝備工程師／屋宇裝備工程師	一級	一級	及格
10	屋宇測量師	助理屋宇測量師／屋宇測量師	一級	一級	及格
11	製圖師	助理製圖師／製圖師	一級	一級	-
12	化驗師	化驗師	一級	一級	及格
13	臨床心理學家（衞生署、入境事務處）	臨床心理學家（衞生署、入境事務處）	一級	一級	-
14	臨床心理學家（懲教署、香港警務處）	臨床心理學家（懲教署、香港警務處）	二級	二級	-
15	臨床心理學家（社會福利署）	臨床心理學家（社會福利署）	二級	二級	及格
16	法庭傳譯主任	法庭二級傳譯主任	二級	二級	及格
17	館長	二級助理館長	二級	二級	-
18	牙科醫生	牙科醫生	一級	一級	-
19	營養科主任	營養科主任	一級	一級	-
20	經濟主任	經濟主任	二級	二級	-
21	教育主任（懲教署）	助理教育主任（懲教署）	一級	一級	-
22	教育主任（教育局、社會福利署）	助理教育主任（教育局、社會福利署）	二級	二級	-
23	教育主任（行政）	助理教育主任（行政）	二級	二級	-
24	機電工程師（機電工程署）	助理機電工程師／機電工程師（機電工程署）	一級	一級	及格
25	機電工程師（創新科技署）	助理機電工程師／機電工程師（創新科技署）	一級	一級	-

	職系	入職職級	英文運用	中文運用	能力傾向測試
26	電機工程師（水務署）	助理電機工程師/ 電機工程師（水務署）	一級	一級	及格
27	電子工程師（民航署、機電工程署）	助理電子工程師/ 電子工程師（民航署、機電工程署）	一級	一級	及格
28	電子工程師（創新科技署）	助理電子工程師/電子工程師(創新科技署)	一級	一級	-
29	工程師	助理工程師/ 工程師	一級	一級	及格
30	娛樂事務管理主任	娛樂事務管理主任	二級	二級	及格
31	環境保護主任	助理環境保護主任/ 環境保護主任	二級	二級	及格
32	產業測量師	助理產業測量師/ 產業測量師	一級	一級	-
33	審查主任	審查主任	二級	二級	及格
34	行政主任	二級行政主任	二級	二級	及格
35	學術主任	學術主任	一級	一級	-
36	漁業主任	助理漁業主任/ 漁業主任	一級	一級	及格
37	警察福利主任	警察助理福利主任	二級	二級	-
38	林務主任	助理林務主任/ 林務主任	一級	一級	及格
39	土力工程師	助理土力工程師/ 土力工程師	一級	一級	及格
40	政府律師	政府律師	二級	一級	-
41	政府車輛事務經理	政府車輛事務經理	一級	一級	-
42	院務主任	二級院務主任	二級	二級	及格
43	新聞主任(美術設計)/(攝影)	助理新聞主任（美術設計）/ （攝影）	一級	一級	-
44	新聞主任（一般工作）	助理新聞主任（一般工作）	二級	二級	及格
45	破產管理主任	二級破產管理主任	二級	二級	及格
46	督學（學位）	助理督學（學位）	二級	二級	-
47	知識產權審查主任	二級知識產權審查主任	二級	二級	及格
48	投資促進主任	投資促進主任	二級	二級	-
49	勞工事務主任	二級助理勞工事務主任	二級	二級	及格
50	土地測量師	助理土地測量師/ 土地測量師	一級	一級	-

	職系	入職職級	英文運用	中文運用	能力傾向測試
51	園境師	助理園境師/ 園境師	一級	一級	及格
52	法律翻譯主任	法律翻譯主任	二級	二級	-
53	法律援助律師	法律援助律師	二級	一級	及格
54	圖書館館長	圖書館助理館長	二級	二級	及格
55	屋宇保養測量師	助理屋宇保養測量師/ 屋宇保養測量師	一級	一級	及格
56	管理參議主任	二級管理參議主任	二級	二級	及格
57	文化工作經理	文化工作副經理	二級	二級	及格
58	機械工程師	助理機械工程師/ 機械工程師	一級	一級	及格
59	醫生	醫生	一級	一級	-
60	職業環境衞生師	助理職業環境衞生師/ 職業環境衞生師	二級	二級	及格
61	法定語文主任	二級法定語文主任	二級	二級	-
62	民航事務主任（民航行政管理）	助理民航事務主任（民航行政管理） 民航事務主任（民航行政管理）	二級	二級	及格
63	防治蟲鼠主任	助理防治蟲鼠主任/ 防治蟲鼠主任	一級	一級	及格
64	藥劑師	藥劑師	一級	一級	-
65	物理學家	物理學家	一級	一級	及格
66	規劃師	助理規劃師/ 規劃師	二級	二級	及格
67	小學學位教師	助理小學學位教師	二級	二級	-
68	工料測量師	助理工料測量師/ 工料測量師	一級	一級	及格
69	規管事務經理	規管事務經理	一級	一級	-
70	科學主任	科學主任	一級	一級	-
71	科學主任（醫務）(衞生署)	科學主任（醫務）（衞生署）	一級	一級	-
72	科學主任（醫務）（食物環境衞生署）	科學主任（醫務）（食物環境衞生署）	一級	一級	及格
73	管理值班工程師	管理值班工程師	一級	一級	-
74	船舶安全主任	船舶安全主任	一級	一級	-
75	即時傳譯主任	即時傳譯主任	二級	二級	-

	職系	入職職級	英文運用	中文運用	能力傾向測試
76	社會工作主任	助理社會工作主任	二級	二級	及格
77	律師	律師	二級	一級	-
78	專責教育主任	二級專責教育主任	二級	二級	-
79	言語治療主任	言語治療主任	一級	一級	-
80	統計師	統計師	二級	二級	及格
81	結構工程師	助理結構工程師/ 結構工程師	一級	一級	及格
82	電訊工程師（香港警務處）	助理電訊工程師/ 電訊工程師（香港警務處）	一級	一級	-
83	電訊工程師（通訊事務管理局辦公室）	助理電訊工程師/ 電訊工程師（（通訊事務管理局辦公室））	一級	一級	及格
84	電訊工程師（香港電台）	高級電訊工程師/ 助理電訊工程師/ 電訊工程師（香港電台）	一級	一級	-
85	電訊工程師（消防處）	高級電訊工程師（消防處）	一級	一級	-
86	城市規劃師	助理城市規劃師/ 城市規劃師	二級	二級	及格
87	貿易主任	二級助理貿易主任	二級	二級	及格
88	訓練主任	二級訓練主任	二級	二級	及格
89	運輸主任	二級運輸主任	二級	二級	及格
90	庫務會計師	庫務會計師	二級	二級	及格
91	物業估價測量師	助理物業估價測量師/ 物業估價測量師	一級	一級	及格
92	水務化驗師	水務化驗師	一級	一級	及格

資料截至2016年3月

12 個最多公務員的部門

部門	實際人數
香港警務處	33650
消防處	10420
食物環境衞生署	10118
康樂及文化事務署	9194
房屋署	8775
入境事務處	7668
懲教署	6601
香港海關	6294
衞生署	6113
社會福利署	5833
郵政署	5204
教育局	5048
其他部門	55425
總數	**170343**

* 統計截至 2018 年 2 月 15 日止

PART

TWO

考試精讀題庫—

I. 演繹推理

通常演繹推理的題目分以下幾種型式出現，只要掌握題目的類型，自然能以推理判斷解答演繹推理。

(1)直接得出結論的類型

【例1】有一段時間，街上的年輕女性都穿著一種高跟的「鬆糕」皮鞋，但這種鞋不美，是男青年的共識，不久這種皮鞋就少見了。如今，在男青年的衣櫃裡，雙排扣的西裝可能已落滿了灰塵，這種西裝氣派、莊重，但有拒年輕女性千里之外的感覺。可見（　）。

A. 女人都愛趕潮流

B. 市場上已經沒有高跟「鬆糕」皮鞋和雙排扣西裝銷售了

C. 穿高跟皮鞋沒有女人味，穿雙排扣西裝男人味又太濃

D. 男人和女人流行哪種服飾，很大程度上取決於異性是否認同

(2)間接得出結論的類型

【例2】甲、乙、丙三人是同一家公司的職員，他們的未婚妻A、B、C也都是這家公司的職員。知情者介紹說：「A的未婚夫是乙的好友，並在三個男子中最年輕；丙的年齡比C的未婚夫大。」依據該知情者提供的資訊，我們可以推出三對夫妻分別是（　）。

A. 甲-A，乙-B，丙-C　　B. 甲-A，乙-C，丙-B

C. 甲-B，乙-C，丙-A　　D. 甲-C，乙-B，丙-A

【例3】據《科學日報》消息，1998年5月，瑞典科學家在有關領域的研究中首次提出，一種對防治老年癡呆症有特殊功效的微量元素，只有在未經加工的加勒比椰果中才能提取。如果《科學日報》的上述消息是真實的，那麼，以下哪項不可能是真實的？

（1）1997年4月，芬蘭科學家在相關領域的研究中提出過，對防治老年癡呆症有特殊功效的微量元素，除了未經加工的加勒比椰果，不可能在其他物件中提取。

（2）荷蘭科學家在相關領域的研究中證明，在未經加工的加勒比椰果中，並不能提取對防治老年癡呆症有特殊功效的微量元素，這種微量元素可以在某些深海微生物中提取。

（3）著名的蘇格蘭醫生查理博士在相關的研究領域中證明，該微量元素對防治老年癡呆並沒出現特殊功效。

A.只有（1） B.只有（2） C.只有（3） D.只有（2）和（3）

【例4】甲、乙、丙和丁是同班同學。甲說：「我班同學都是團員。」乙說：「丁不是團員。」丙說：「我班有人不是團員。」丁說：「乙也不是團員。」

已知只有一個說假話，則可推出以下哪項斷定是真的？

A. 説假話的是甲，乙不是團員

B. 説假話的是乙，丙不是團員

C. 説假話的是丙，丁不是團員

D. 説假話的是丁，乙不是團員

(3)邏輯推理類

【例5】所有的詩人都是文學家，有的文學家是詩人，張中是文學家，則下列選擇正確的是（　）。

A. 張中是詩人

B. 張中不是詩人

C. 張中可能是詩人

D. 張中不是文學家就是詩人

【例6】以「如果甲乙都不是作案者，那麼丙是作案者」為一前提，若再增加另一前提可必然推出「乙是作案者」的結論。下列哪項最適合作這一前提？（　）

A. 丙是作案者

B. 丙不是作案者

C. 甲不是作案者

D. 甲和丙都不是作案者

答案

(1)： (例1) D

(2)： (例2) B　(例3) A　(例4) A

(3)： (例5) C　(例6) D

假定短文的內容都是正確，選出一個或一組推論。

1. 今年M市開展了一次前所未有的化妝品廣告大戰。但是調查表明，只有 25%的M市居民實際使用化妝品。這說明化妝品公司的廣告投入有很大的盲目性。

   以下哪項陳述最有力地加強了上述結論？（　　）

   A. 化妝品公司做廣告是因為產品供過於求

   B. 去年實際使用化妝品M市居民有30％

   C. 大多數不使用化妝品的居民不關心其廣告宣傳

   D. 正是因為有25％的居民使用化妝品，才要針對他們做廣告

2. 第二次世界大戰期間，海洋上航行的商船常常遭到轟炸機的襲擊，許多商船都先後在船上架設了高射炮。但是，商船在海上搖晃得比較厲害，用高射炮射擊天上的飛機是很難命中的。戰爭結束後，研究人員發現，從整個戰爭期間架設過高射炮的商船的統計資料看，擊落敵機的命中率只有4%。因此，研究人員認為，商船上架設高射炮是得不償失的。

   以下哪個如果為真，最能削弱上述研究人員的論？（　　）

A. 在戰爭期間，為架設高射炮的商船，被擊沉的比例高達25％；而架設了高射炮的商船，被擊沉的比例只有不到10％

B. 架設了高射炮的商船，即使不能將敵機擊中，在某些情況下也可能將敵機嚇跑

C. 架設高射炮的費用是一筆不小的投入，而且在戰爭結束後，為了運行的效率，還要在花費資金將高射炮拆除。

D. 一般來説，上述商船用於高射炮的費用，只占整個商船的總價值的極少部分

3. 政府應該不允許煙草公司在其營業收入中扣除廣告費用。這樣的話，煙草公司將會繳納更多的稅金。煙草公司只好提高自己的產品價格，而產品價格的提高正好可以起到減少煙草購買的作用。

以下哪項是題目論點的前提？（　　）

A. 煙草公司不可能降低其他方面的成本來抵消多繳的稅金

B. 如果它們需要付高額的稅金，煙草公司將不再繼續做廣告

C. 如果煙草公司不做廣告，香煙的銷售量將受到很大影響

D. 煙草公司由此所增加的稅金應該等於價格上漲所增加的盈利

4. 以前有幾項研究表明，食用巧克力會增加食用者患心臟病的可能性。而一項最新的、更為可靠的研究得出的結論是：食用巧克力與心臟病發病率無關。估計這項研究成果公布以後，巧克力的消費量將會大大增加。

上述推論基於以下哪項假設？（　　）

A. 盡管有些人知道食用巧克力會增加患心臟病的可能性，卻照樣大吃特吃

B. 人們從來也不相信進食巧克力會更容易患心臟病的說法

C. 現在許多人吃巧克力是因為他們沒有聽過巧克力會導致心臟病的說法

D. 現在許多人不吃巧克力完全是因為他們相信巧克力會誘發心臟病

5. 在評獎會上，A、B、C、D、E、F、G、H競爭一項金獎。由一個專家小組投票，票數最多的將獲金獎。

如果A的票數多於B，並且C的票數多於D，那麼E將獲得金獎。

如果B的票數多於A，或者F的票數多於G，那麼H將獲得金獎。

如果D的票數多於C，那麼F將獲得金獎。

如果上述斷定都是真的，並且事實上C的票數多於D，並且E並沒有獲得金獎，以下哪項一定是真的？（　　）

A. H獲獎　　B. F的票數多於G

C. A的票數不比B多　　D. B的票數不比F多

6. 某珠寶商店失竊，甲、乙、丙、丁四人涉嫌被拘審。四人的口供如下：甲：案犯是丙。乙：丁是案犯。丙：如果我作案，那麼丁是主犯。丁：作案的不是我。四個口供中只有一個是假的。

如果以上斷定為真，則以下哪項是真的？（　　）

A. 説假話的是甲，作案的是乙

B. 説假話的是丁，作案的是丙和丁

C. 説假話的是乙，作案的是丙

D. 説假話的是丙，作案的是丙

7. 亞里士多德學院的門口豎著一塊牌子，上面寫著"不懂邏輯者不得入內"。這天，來了一群人，他們都是懂邏輯的人。如果牌子上的話得到準確的理解和嚴格的執行，那麼以下諸斷定中，只有一項是真的。這一真的斷定是（　　）。

A. 他們可能不會被允許進入

B. 他們一定不會被允許進入

C. 他們一定會被允許進入

D. 他們不可能被允許進入

8. 糧食可以在收割前在期貨市場進行交易。如果預測水稻產量不足，水稻期貨價格就會上升；如果預測水稻豐收，水稻期貨價格就會下降。假設今天早上，氣像學家們預測從明天開始水稻產區會有適量降雨。因為充分的潮濕對目前水稻的生長非常重要，所以今天的水稻期貨價格會大幅下降。下面哪項如果正確，最嚴重地削弱以上的觀點？

A. 農業專家們今天宣布，一種水稻病菌正在傳播

B. 本季度水稻期貨價格的波動比上季度更加劇烈

C. 氣像學家們預測的明天的降雨估計很可能會延伸到谷物產區以外

D. 在關鍵的授粉階段沒有接受足夠潮濕的穀物不會取得豐收

9. 發達國家中冠心病的發病率大約是發展中國家的三倍。有人認為，這主要歸咎於發達國家中人們的高脂肪、高蛋白、高熱量的食物攝入。相對來說，發展中國家較少有人具備生這種“富貴病”的條件。其實，這種看法很難成立。因為，目前發達國家的人均壽命高於70歲，而發展中國家的人均壽命還不到50歲。

以下哪項如果成立，最能加強上述反駁？（　　）

A. 統計資料顯示，冠心病患者相對集中在中老年年齡層，即45歲以上

B. 目前冠心病患者呈年輕化趨勢

C. 發展中國家人們的高脂肪、高蛋白、高熱量食物的攝入量，無論是總量還是人均量，都在逐年增長

D. 相對發展中國家來說，發達國家的人們具有較高的防治冠心病的常識和較好的醫療條件

10. 有時為了醫治一些危重病人，醫院允許使用海洛英作為止痛藥。其實，這樣做是應當禁止的。因為，毒品販子會通過這種渠道獲取海洛英，對社會造成嚴重危害。

以下哪個如果為真，最能削弱以上的論證？（　　）

A. 有些止痛藥可以起到和海洛因一樣的止痛效果

B. 用於止痛的海洛英在數量上與用於做非法交易的比起來是微不足道的

C. 海洛英如果用量過大就會致死

D. 在治療過程中，海洛英的使用不會使病人上癮

11. 雖然菠菜中含有豐富的鈣，但同時含有大量的漿草酸，漿草酸會有力地阻止人體對鈣的吸收。因此，一個人要想攝入足夠的鈣，就必須用其他含鈣豐富的食物來取代菠菜。

以下哪個如果為真，最能削弱題目的論證？（　　）

A. 大米中不含鈣，但含有中和漿草酸並改變其性能的城性物質

B. 奶製品中的鈣含量要高於菠菜，許多經常食用菠菜的人也食用奶製品

C. 在人的日常飲食中，除了菠菜以外，事實上大量的蔬菜都含有鈣

D. 菠菜中除了鈣以外，還含有其他豐富的營養素；另外，漿草酸只阻止人體對鈣的吸收，並不阻止其他營養的吸收

12. 全國各地的電話公司目前開始為消費者提供電子接線員系統，然而，在近期內，人工接線員並不會因此減少。

除了下列哪項外，其他各項均有助於解釋上述現像？（　　）

A. 需要接線員幫助的電話數量劇增

B. 儘管已經過測試，新的電子接線員系統要全面發揮功能還需進一步調整

C. 如果在目前的合同期內解雇人工接線員，有關方面將負法律責任

D. 新的電子接線員的工作效率兩倍於人工接線員

13. 《能源效益標籤計劃》規定，能效五級是最低的能效標準，是產品上市的最低要求，低於這個要求就不許生產銷售。而節能標籤和能效標籤是兩個不同的概念。目前節能空調和節能冰箱的認證標準是能效二級，所有的節能產品必須達到二級能效標準以上。但這也並不是說所有標有二級或一級能效標籤的產品就是節能產品，這樣的產品只有再經過認證才能決定是否屬於節能產品。根據以上信息，下列結論正確的是：

A. 節能產品肯定標有二級或一級能效標籤

B. 所有貼有能效標籤的產品都是節能產品

C. 達到二級能效標準就可以認為是節能產品了

D. 能效五級的產品是質量合格的產品，也是節能產品

14. 有報告指出，今年上半年，國內手機累計銷售超過6000萬部，其中國產品牌手機共銷售2800萬部。因此，有媒體判斷國產手機復蘇了。以下哪一選項如果為真，將有力地支持上述判斷？

A. 手機銷量統計不包括水貨手機，而水貨手機的銷量巨大

B. 今年上半年，國家採取措施，限制國外品牌手機進入中國市場

C. 今年下半年，國產手機銷量遠高於其他品牌手機，並繼續保持這一勢頭

D. 手機銷量是依據進網許可證發放數量來統計的，但這些手機可能並未全部進入用戶手中

15. 很多人以為只有抽煙的老人才會得肺癌，但某國一項最新的統計顯示：近年來該國肺癌導致的女性死亡人數比乳腺癌、子宮內膜癌和卵巢癌三種癌症加起來還多，而絕大多數的婦女們根本沒有意識到這一點。由此無法推出的是：

A. 肺癌是導致該國人口死亡的首要原因，應當得到極大的重視

B. 普遍認為男性比女性更容易患肺癌的觀點，可能是片面的

C. 煙草並不是肺癌的惟一致病源，還有很多因素也參與到肺癌的發病過程中

D. 肺癌未引起廣大女性的重視，是因為她們認為自己不抽煙，不可能得肺癌

16. 一般病菌多在室溫環境生長繁殖，低溫環境停止生長，僅能維持生命，而耶爾森氏菌卻恰恰相反，不但不怕低溫寒冷，而且只有在0℃左右才大量繁殖。冰箱裡存儲的食物，使耶爾森氏菌處於最佳生長狀態。由此可以推出：

A. 耶爾森氏菌在室溫環境無法生存

B. 一般病菌生長的環境也適合耶爾森氏菌生長

C. 耶爾森氏菌的最佳生長溫度不適合一般病菌

D. 0℃環境下，冰箱裡僅存在耶爾森氏菌

17. 在同一側的房號為1、2、3、4的四間房裡，分別住著來自韓國、法國、英國和德國的四位專家。有一位記者前來採訪他們：

①韓國人說：「我的房號大於德國人，且我不會說外語，也無法和鄰居交流」；

②法國人說：「我會說德語，但我卻無法和我的鄰居交流」；

③英國人說：「我會說韓語，但我只可以和一個鄰居交流」；

④德國人說：「我會說我們這四個國家的語言」。

那麼，按照房號從小往大排，房間裡住的人的國籍依次是（ ）。

A. 英國 德國 韓國 法國

B. 法國 英國 德國 韓國

C. 德國 英國 法國 韓國

D. 德國 英國 韓國 法國

18. 某律師事務所共有12名工作人員。(1)有人會使用計算機；(2)有人不會使用計算機；(3)所長不會使用計算機。這三個命題中只有一個是真的，以下哪項正確地表示了該律師事務所會使用計算機的人數？（ ）

A. 12人都會使用　B. 12人沒人會使用

C. 僅有一人會使用　D. 不能確定

19. 規定汽車必須裝安全帶的制度是為了減少車禍傷亡，但在安全帶保護下，司機將車開得更快，事故反而增加了。司機有安全帶保護，自身傷亡減少了，而路人傷亡增加了。這一事實表明：

A. 對實施效果考慮不周的制度往往事與願違

B. 安全帶制度必須與嚴格限速的制度同時出台

C. 汽車裝安全帶是通過犧牲路人利益來保護司機的措施

D. 制度在產生合意結果的同時也會產生不合意的結果

20. 航天局認為優秀宇航員應具備三個條件：第一，豐富的知識；第二，熟練的技術；第三，堅強的意志。現有至少符合條件之一的甲、乙、丙、丁四位優秀飛行員報名參選，已知：

①甲、乙意志堅強程度相同；

②乙、丙知識水平相當；

③丙、丁並非都是知識豐富；

④四人中三人知識豐富、兩人意志堅強、一人技術熟練。

航天局經過考察，發現其中只有一人完全符合優秀宇航員的全部條件。他是：

A. 甲　B. 乙　C. 丙　D. 丁

21. 遇到高溫時，房屋建築材料會發出獨特的聲音。聲音感應報警器能夠精確探測這些聲音，提供一個房屋起火的早期警報，使居住者能在被煙霧困住之前逃離。由於煙熏是房屋火災人員傷亡最通常的致命因素，所以安裝聲音感應報警器將會有效地降低房屋火災的人員傷亡。下列哪一個假設如果正確，最能反駁上面的論述？

A. 聲音感應報警器廣泛使用的話，其高昂成本將下降

B. 在完全燃燒時，許多房屋建築材料發出的聲音在幾百米外也可聽見

C. 許多火災開始於室內的沙發座墊或床墊，產生大量煙霧卻不發出聲音

D. 在一些較大的房屋中，需要多個聲音感應報警器以達到足夠的保護

22. 在農業發展初期，很少遇到昆蟲問題。這一問題是隨著農業的發展而產生的——在大面積土地上僅種一種穀物，這樣的種植方法為某些昆蟲的猛增提供了有利條件。很明顯，一種食麥昆蟲在專種麥子的農田裡比在其他農田裡繁殖起來要快得多。上述論斷不能解釋下列哪種情況？

A. 一種由甲蟲帶來的疾病掃蕩了某城市街道兩旁的梧桐樹

B. 控制某一種類生物的棲息地的適宜面積符合自然發展規律的格局

C. 遷移到新地區的物種由於逃離了其天敵對它的控制而蓬勃發展起來

D. 楊樹的害蟲在與其他樹木摻雜混種的楊樹林中的繁殖速度會受到限制

23. 一份關於酸雨的報告總結說，「大多數森林沒有被酸雨損害。」而反對者堅持應總結為，「大多數森林沒有顯示出明顯的被酸雨損害的症狀，如不正常的落葉。生長速度的減慢或者更高的死亡率。」下面哪項如果正確，最能支持反對者的觀點？

A. 目前該地區的一些森林正在被酸雨損害

B. 酸雨造成的損害程度在不同森林之間具有差異

C. 酸雨可能正在造成症狀尚未明顯的損害

D. 報告沒有把酸雨對此地區森林的損害與其他地區相比較

24. 維生素 E 是抗氧化劑，能夠清除體內的自由基。於是，保健品商家把維生素E作為提高免疫力、抗癌、抗衰老的靈丹妙藥來宣傳。科學家通過實驗發現：如果食物中維生素E的含量為每毫升5微克，能顯著延長果蠅的壽命，但是如果維生素E的含量增加到每毫升25微克，果蠅的壽命反而縮短了。其實，細胞中的自由基參與了許多重要的生命活動，比如細胞增殖、細胞間通訊、細胞凋亡、免疫反應等。由此推論不正確的是：

A. 自由基有其獨特的作用，對機體而言是不可或缺的

B. 科學家對果蠅的實驗揭示了「過猶不及」的道理

C. 維生素 E 的含量超過25微克時，會危及到人的生命

D. 維生素是維持人體生命的必要物質，但過量服用時也會威脅生命

25. 未來深海水下線纜的外皮將由玻璃製成，而不是特殊的鋼材或鋁合金。因為金屬具有顆粒狀的微觀結構，在深海壓力之下，粒子交界處的金屬外皮容易斷裂。而玻璃看起來雖然是固體，但在壓力之下可以流動，因此可以視為液體。由此可以推出：

A. 玻璃沒有顆粒狀的微觀結構

B. 一切固體幾乎都可以被視為緩慢流動的液體

C. 玻璃比起鋼材或鋁合金，更適合做建築材料

D. 與鋼材相比，玻璃的顆粒狀的微觀結構流動性更好

26. 一家飛機發動機製造商開發出了一種新型發動機，安全性能要好於舊型發動機。在新舊兩種型號的發動機同時被銷售的第一年，舊型發動機的銷量超過了新型發動機，該製造商於是得出結論認為安全性並非客戶的首要考慮。下面哪項如果正確，會最嚴重地削弱該製造商的結論？

A. 新型發動機和舊型發動機沒有特別大的價格差別

B. 新型發動機可以被所有的使用舊型發動機的飛機使用

C. 私人飛機主和航空公司都從這家飛機發動機製造商這裡購買發動機

D. 客戶認為舊型發動機在安全性方面比新型號好，因為他們對舊型發動機的安全性了解更多

27. 在就業者中存在一種「多元的幻覺」：認為在這個多元開放的時代，每個人對自己的未來負責，對未來之路的選擇是多元的，自由的。但看看現實就知道，這種選擇下的目標指向是一元的，大家都一窩蜂地流向了城市，盯住了高薪白領職位，以為是個性選擇，實際都彙合進同一條河流；以為是多元，實際被同化為一元；以為是自由的追求，實際都被一種封閉的思想禁錮——這便是「多元的幻覺」。由此可以推出的是：

A. 高薪職位的競爭將更加激烈

B. 多元的選擇客觀上是不存在的

C. 就業者實際上沒有自由選擇的權利

D. 社會並沒有給就業者提供多元的選擇

答案與解釋

1. 答案： C。廣告投入不針對有效的市場需求就是盲目投入，而A、B、D項都不能説明廣告投入的有效性問題，只有C項正確。

2. 答案： A。A項説明商船上自設高射炮是有效的，不是得不償失的。

3. 答案： A。A項中“煙草公司可能降低其他方面的成本來抵消多繳的稅金”是題目中提高產品售價的原因。

4. 答案： D。本題是問在什麼情況下“食用巧克力與心臟病發病率無關……巧克力的消費量將會大大增加”，據此只有D項正確。

5. 答案： C。由“事實上C的票數多於D，並且E並沒有獲得金獎”可知A的票數不比B多。

6. 答案： B。從題目可知乙的口供與丁的口供相互矛盾，由此知其中一人必説假話，也就是説甲、丙説的都是真話，即案犯是丙且丁是主犯，丁説假話。

7. 答案： A。依據題目可知不懂邏輯者不得入內，但這並不必然推出懂邏輯者就能入內，因此，A為正確答案。

8. 答案： A。B、C是無關項，可排除。D項是支持題幹中的觀點，故也排除。

9. 答案： A。A項中説冠心病患者相對集在中45歲以上的人群中，而題目中發達國家的人均壽命高於70歲，而發展中國家的人均壽命還不到50歲。故A正確。

10. 答案： B。若用於止痛的海洛英在數量上與用做非法交易的海洛英比起來是微不足道的，則其不可能對社會造成嚴重危害。故選B。

11. 答案： A。若想削弱上述論證，只要證明食用菠菜同樣能攝入足量的鈣即可，A項中吃大米會中和菠菜中的漿草酸而不影響人體對菠菜中鈣的攝入。其他三項都是加強題目中的論述，而不是削弱。

12. 答案： D。D項無助於解釋題目中所述的現象，相反，其是在削弱“人工接線員並不會因此而減少”的論述。

13. 答案： A。此種類型的題目需要嚴格依照題目所給的條件來進行演繹推理，因此認真分析、理解題目是解題的關鍵。本題中，B、C、D都不能直接從題目的陳述中直接推出，故選A。

14. 答案： C。此題比較簡單，顯然，只有C能支持題目中的判斷。

15. 答案： A。本題是選非題。題目中只是説該國肺癌導致的女性死亡人數比乳線癌、子宮內膜癌和卵巢癌三種癌症加起來裏多，據此並不能得出肺癌是該國人口死亡的首要原因，故選A。

16. 答案： C。題目說一般病菌在但溫環境中停止生長，僅能維持生命。而耶爾森菌則在0℃時才大量繁殖，故可以直接推出C項的內容。

17. 答案： C。根據英國人和德國人所說，可推知英國人不能同時和韓國人和德國人相鄰，故排除D；又根據韓國人和德國人所說，可推知德國人和韓國人不能相鄰，故排除A、B，故選C。

18. 答案： A。用推斷法解此題，據題目敍述可知三個命題只有一個為真；那麼將選項帶入三個命題中去分析可知只有A項正確。

19. 答案： D。與「自身傷亡減少了，而路人傷亡增加了」相對應的是「制度在產生合意結果的同時也會產生不意的結果」，故選D。

20. 答案： C。根據②③④可推知丁知識不豐富，故排除D，但是根據這個推論咎題目所給條件並不能直接得出結論。這時，我們可以採用假設法。先假設甲符合所有條件，那麼根據①④可知甲、乙意志堅強，所以丁不具備意志堅強的條件，同時，根據④可知丁也不具備技術熟練的條件，這件丁三個條件都不具備，這顯然和題「至少符合條件之一」相矛盾，故先前假設錯誤，故又排除B，所以選C。

21. 答案： C。注意本題要求的是最佳答案，盡管C、D都符合題意，但C是最佳答案。

22. 答案： C。從題目的陳述中可以得出的中心結論是：大面積土地上種植單一作物有利於昆蟲的繁殖。A、B、D都與此論斷相符，故選C。

23. 答案： C。本題要求選擇能夠支持反對者的觀點的選項。B、D是無關項，可以排除。比較A、C，因為題目中反對者的觀點是大多數森林沒有顯示明顯的被酸雨損害的症狀，故選C。

24. 答案： C。本題也是選非題。題目中並沒有說維生素E的含量超過25微克時，就會危及到人的生命，又只是說對果蠅有影響。故選C。

25. 答案： A。根據題意，金屬具有顆粒狀的微觀結構，所以在深海壓力之下，粒子交界處的金屬外皮容易斷裂。而玻璃可以視為液體，在壓力之下可以流動而不斷裂。那麼，其肯定不具有顆粒狀的微觀結構，故選A。

26. 答案： D。A、B、C都從不同角度支持了製造商的結論，只有D項嚴重地削弱製造商的結論。

27. 答案： B。題目中的核心觀點是「多元的幻覺」；盡管選擇的對象是對元的，但是選擇的結果卻是一元的。所以，從結果來看，所謂多元只是一種並不存在的幻覺而已，故選B。

II. Verbal Reasoning

This test includes a number of short passages of text followed by statements based on the information given in the passage. You are asked to indicate whether the statements are (A) True or (B) False, or whether it is (C) Can't Tell. In answering these questions, use only the information given in the passage and do not try and answer them in the light of any more detailed knowledge which you personally may have.

Assume the passage is true and decide whether the statements are either: (A) True, (B) False or (C) Can' t Tell.

Passage 1 (Question 1 to 4)

Abdominal pain in children may be a symptom of emotional disturbance, especially where it appears in conjunction with phobias or sleep disorders such as nightmares or sleep-walking. It may also be linked to eating habits: a study carried out in the USA found that children with pain tended to be more fussy about what and how much they ate, and to have over-anxious parents who spent a considerable time trying to persuade them to eat. Although abdominal pain had previously been linked to excessive milk-drinking, this research found that children with pain drank rather less milk than those in the control group.

1. There is no clear cause for abdominal pain in children.

2. Abdominal pain in children may be psychosomatic in nature.

3. Drinking milk may help to prevent abdominal pain in children.

4. Children who have problems sleeping are more likely to suffer from abdominal pain.

Passage 2 (Question 5 to 8)

The London congestion charge is aimed at encouraging people to think again about using their vehicles in central London and to choose other forms of transport. Motorists who still wish to travel in or through central London have to pay the daily fee. The scheme is policed by cameras on roads within the congestion zone which read car registration plates. The charge is currently active between 7am and 6.00pm, Monday to Friday, excluding Public Holidays.

The boundary of the zone is formed by the Inner Ring Road, on which there is no charge to drive. All vehicles driving across the £8 charge zone have to pay. However, some vehicles are exempt from the charge (for example, taxis, licensed minicabs, emergency services, blue/orange badge holders, alternative energy vehicles). Others entitled to a discount are mainly residents, but also some vehicle breakdown services as well as selected households living on the zone's border.

5. The congestion charge does not apply to emergency services.

6. The congestion charge does not apply on Public holidays.

7. There is a charge to drive on the Inner Ring Road.

8. If you live on the border of the congestion charge zone, you will be entitled to discount.

Passage 3 (Question 9 to 12)

When Christianity was first established by law, a corrupt form of Latin had become the common language of all the western parts of Europe. The service of the Church accordingly, and the translation of the Bible which was read in churches, were both in that corrupted Latin which was the common language of the country. After the fall of the Roman Empire, Latin gradually ceased to be the language of any part of Europe. However, although Latin was no longer understood anywhere by the great body of the people, Church services still continued to be performed in that language. Two different languages were thus established in Europe: a language of the priests and a language of the people.

9. After the fall of the Roman Empire, people who had previously spoken Latin returned to their original languages.

10. Latin continued to be used in church services because of the continuing influence of Rome.

11. Priests spoke a different language from the common people.

12. Prior to the fall of the Roman Empire, Latin had been established by law as the language of the Church in Western Europe.

Passage 4 (Question 13 to 16)

Purchasing tickets

- Customers boarding at main stations will not be permitted to pass the ticket barrier or board any train without first purchasing a valid ticket for their journey.
- Weekly and Monthly Ticket holders can purchase their tickets up to 4 days in advance of the expiry date of their current ticket. By doing this you can avoid the queues on Monday mornings and at the start of the month.
- Please be aware that once the train doors have closed station staff have no way of gaining access to the train and the ticket barriers will be closed.
- To allow all passengers to board the train safely, ticket barriers at main stations will close 1 minute prior to departure time.

13. Passengers are able to purchase a ticket on board a train at main stations.

14. Monthly ticket holders are unable to purchase their tickets 5 days in advance of the expiry date on their current ticket.

15. More people choose to renew their tickets in advance than those who queue on a Monday morning.

16. Once the train doors close, staff have no access to the train.

17. Guitar strings often go "dead"—become less responsive and bright in tone—after a few weeks of intense use. A researcher whose son is a classical guitarist hypothesized that dirt and oil, rather than changes in the material properties of the string, were responsible. Which of the following investigations is most likely to yield significant information that would help to evaluate the researcher's hypothesis?

A. Determining if a metal alloy is used to make the strings used by classical guitarists
B. Determining whether classical guitarists make their strings go dead faster than do folk guitarists
C. Determining whether identical lengths of string, of the same gauge, go dead at different rates when strung on various brands of guitars
D. Determining whether a dead string and a new string produce different qualities of sound
E. Determining whether smearing various substances on new guitar strings causes them to go dead

18. A milepost on the towpath read "21" on the side facing the hiker as she approached it and "23" on its back. She reasoned that the next milepost forward on the path would indicate that she was halfway between one end of the path and the other. However, the milepost one mile further on read "20" facing her and "24" behind. Which of the following, if true, would explain the discrepancy described above?

A. The numbers on the next milepost had been reversed.
B. The numbers on the mileposts indicate kilometers, not miles.
C. The facing numbers indicate miles to the end of the path, not miles from the beginning.
D. A milepost was missing between the two the hiker encountered.
E. The mileposts had originally been put in place for the use of mountain bikers, not for hikers.

19. Most consumers do not get much use out of the sports equipment they purchase. For example, seventeen percent of the adults in the United States own jogging shoes, but only forty-five percent of the owners jog more than once a year, and only seventeen percent jog more than once a week. Which of the following, if true, casts most doubt on the claim that most consumers get little use out of the sports equipment they purchase?

A. Joggers are most susceptible to sports injuries during the first six months in which they jog.
B. Joggers often exaggerate the frequency with which they jog in surveys designed to elicit such information.
C. Many consumers purchase jogging shoes for use in activities other than jogging.
D. Consumers who take up jogging often purchase an athletic shoe that can be used in other sports.
E. Joggers who jog more than once a week are often active participants in other sports as well.

20. Airline: Newly developed collision-avoidance systems, although not fully tested to discover potential malfunctions, must be installed immediately in passenger planes. Their mechanical warnings enable pilots to avoid crashes. Pilots: Pilots will not fly in planes with collision-avoidance systems that are not fully tested. Malfunctioning systems could mislead pilots, causing crashes. The pilots' objection is most strengthened if which of the following is true?

A. It is always possible for mechanical devices to malfunction.

B. Jet engines, although not fully tested when first put into use, have achieved exemplary performance and safety records.

C. Although collision-avoidance systems will enable pilots to avoid some crashes, the likely malfunctions of the not-fully-tested systems will cause even more crashes.

D. Many airline collisions are caused in part by the exhaustion of overworked pilots.

E. Collision-avoidance systems, at this stage of development, appear to have worked better in passenger planes than in cargo planes during experimental flights made over a six-month period.

ANSWERS

1. A (True): There is no clear cause for abdominal pain in children.

2. A (True): Abdominal pain in children may be psychosomatic in nature.

3. C (Can't tell): Drinking milk may help to prevent abdominal pain in children.

4. C (Can't tell): Children who have problems sleeping are more likely to suffer from abdominal pain.

5. A (True)

6. A (True)

7. B (False)

8. C (Can't tell)

9. A (True): After the fall of the Roman Empire, people who had previously spoken Latin returned to their original languages.

10. B (False): Latin continued to be used in church services because of the continuing influence of Rome.

11. A (True): Priests spoke a different language from the common people.

12. B (False): Prior to the fall of the Roman Empire, Latin had been established by law as the language of the Church in Western Europe.

13. B (False)

14. A (True)

15. C (Can't tell)

16. A (True)

17. E

18. C

19. C

20. C

III. Data Sufficiency Test

Get to know the test. Though the test only covers arithmetic, algebra, geometry, and word problems, you will need to become familiar with how the questions are asked.

Math Concepts You Should Know

The data sufficiency questions cover math that nearly any college-bound high school student will know. In addition to basic arithmetic, you can expect questions testing your knowledge of averages, fractions, decimals, algebra, factoring, and basic principles of geometry such as triangles, circles, and how to determine the areas and volumes of simple geometric shapes.

The Answer Choices

The following questions will all have the exact same answer choices. The answer choices are summarized below as you will see them on the exam.

Statement 1 alone is sufficient but statement 2 alone is not sufficient to answer the question asked.

Statement 2 alone is sufficient but statement 1 alone is not sufficient to answer the question asked.

Both statements 1 and 2 together are sufficient to answer the question but neither statement is sufficient alone.

Each statement alone is sufficient to answer the question.

Statements 1 and 2 are not sufficient to answer the question asked and additional data is needed to answer the statements.

Use Process of Elimination

If statement 1 is insufficient, then choices A and D can immediately be eliminated.

Similarly, if statement 2 is insufficient, then choices B and D can immediately be eliminated.

If either statement 1 or 2 is sufficient on its own, then choices C and E can be eliminated.

A Simple 4 Step Process for Answering These Questions

Many test takers make the mistake of not arming themselves with a systematic method for analyzing the answer choices for these questions. Overlooking even one step in the process outlined below can make a big difference in the final quantitative score you will be reporting to your selected business schools.

1.) Study the questions carefully. The questions generally ask for one of 3 things: 1) a specific value, 2) a range of numbers, or 3) a true/false value. Make sure you know what the question is asking.

2.) Determine what information is needed to solve the problem. This will, obviously, vary depending on what type of question is being asked. For example, to determine the area of a circle, you need to know the circle's diameter, radius, or circumference. Whether or not statements 1 and/or 2 provide that information will determine which answer you choose for a data sufficiency question about the area of a circle.

3.) Look at each of the two statements independently of the other. Follow the process of elimination rules covered above to consider each statement individually.

4.) If step 3 did not produce an answer, then combine the two statements.

If the two statements combined can answer the question, then the answer choice is C. Otherwise, E.

Data Sufficiency Tips and Strategies

Use only the information given in the questions. Do not rely on a visual assessment of a diagram accompanying a geometry question to determine angle sizes, parallel lines, etc. In addition, do not carry any information over from one question to the next. Each question in the data sufficiency section stands on its own. You can count on seeing at least a few questions where a wrong answer choice tries to capitalize on this common fallacy.

Do not get bogged down with complicated or lengthy calculations. As we stated before, these questions are designed to test your ability to think conceptually, not to solve math problems.

Use process of elimination. If time becomes an issue, you can always look at the 2 statements in either order. Remember, the order you analyze the two statements in doesn't matter, so long as you begin by looking at them individually. If you find statement 1 confusing, you can save time by skipping to statement 2 and seeing whether it can help you eliminate incorrect answer choices.

Be on the lookout for statements that tell you the same thing in different words. When the 2 statements convey the same exact information, you will know, through process of elimination, that the correct answer choice is either D or E. A favorite ploy of testers is to mix ratios and percentages. Here is an example where Statement 2 simply states backwards the exact same information provided by Statement 1.

1 *x is 50% of y*

2. *the ratio of y:x is 2:1*

Make real-world assumptions where necessary. You must assume that, in certain abstract questions such as "What is the value of x?", that x might be a fraction and/or a negative number.

Directions: Each question below is followed by 2 statements numbered (1) and (2). The questions have to be answered in terms of choices A to E. Mark your answer choice as;

If Statement (1) ALONE is sufficient but Statement (2) ALONE is not sufficient.

If Statement (2) ALONE is sufficient but Statement (1) ALONE is not sufficient.

If BOTH Statements TOGETHER are sufficient, but NEITHER Statement alone is sufficient.

If Each Statement ALONE is sufficient.

If Statements (1) and (2) TOGETHER are NOT sufficient.

Choose a combination of clues to solve the problem.

1. How many ewes (female sheep) in a flock of 50 sheep are black?

   There are 10 rams (male sheep) in the flock.

   Forty percent of the animals are black.

   A. statement 1 alone is sufficient, but statement 2 alone is not sufficient to answer the question
   B. statement 2 alone is sufficient, but statement 1 alone is not sufficient to answer the question
   C. both statements taken together are sufficient to answer the question, but neither statement alone is sufficient
   D. each statement alone is sufficient
   E. statements 1 and 2 together are not sufficient, and additional data is needed to answer the question

2. Is the length of a side of equilateral triangle E less than the length of a side of square F?

   The perimeter of E and the perimeter of F are equal.

   The ratio of the height of triangle E to the diagonal of square F is $2\sqrt{3} : 3\sqrt{2}$.

   A. statement 1 alone is sufficient, but statement 2 alone is not sufficient to answer the question

   B. statement 2 alone is sufficient, but statement 1 alone is not sufficient to answer the question

   C. both statements taken together are sufficient to answer the question, but neither statement alone is sufficient

   D. each statement alone is sufficient

   E. statements 1 and 2 together are not sufficient, and additional data is needed to answer the question

3. If a and b are both positive, what percent of b is a?

$a = 3/11$

$b/a = 20$

A. statement 1 alone is sufficient, but statement 2 alone is not sufficient to answer the question

B. statement 2 alone is sufficient, but statement 1 alone is not sufficient to answer the question

C. both statements taken together are sufficient to answer the question, but neither statement alone is sufficient

D. each statement alone is sufficient

E. statements 1 and 2 together are not sufficient, and additional data is needed to answer the question

4. A wheel of radius 2 meters is turning at a constant speed. How many revolutions does it make in time T?

   T = 20 minutes.

   The speed at which a point on the circumference of the wheel is moving is 3 meters per minute.

   A. statement 1 alone is sufficient, but statement 2 alone is not sufficient to answer the question

   B. statement 2 alone is sufficient, but statement 1 alone is not sufficient to answer the question

   C. both statements taken together are sufficient to answer the question, but neither statement alone is sufficient

   D. each statement alone is sufficient

   E. statements 1 and 2 together are not sufficient, and additional data is needed to answer the question

5. Are the integers x, y and z consecutive?

The arithmetic mean (average) of x, y and z is y.

$y-x = z-y$

A. statement 1 alone is sufficient, but statement 2 alone is not sufficient to answer the question

B. statement 2 alone is sufficient, but statement 1 alone is not sufficient to answer the question

C. both statements taken together are sufficient to answer the question, but neither statement alone is sufficient

D. each statement alone is sufficient

E. statements 1 and 2 together are not sufficient, and additional data is needed to answer the question

6. Is $x > 0$?

   $-2x < 0$

   $x^3 > 0$

   A. statement 1 alone is sufficient, but statement 2 alone is not sufficient to answer the question

   B. statement 2 alone is sufficient, but statement 1 alone is not sufficient to answer the question

   C. both statements taken together are sufficient to answer the question, but neither statement alone is sufficient

   D. each statement alone is sufficient

   E. statements 1 and 2 together are not sufficient, and additional data is needed to answer the question

7. A certain straight corridor has four doors, A, B, C and D (in that order) leading off from the same side. How far apart are doors B and C?

The distance between doors B and D is 10 meters.

The distance between A and C is 12 meters.

A. statement 1 alone is sufficient, but statement 2 alone is not sufficient to answer the question

B. statement 2 alone is sufficient, but statement 1 alone is not sufficient to answer the question

C. both statements taken together are sufficient to answer the question, but neither statement alone is sufficient

D. each statement alone is sufficient

E. statements 1 and 2 together are not sufficient, and additional data is needed to answer the question

8. Given that x and y are real numbers, what is the value of x + y ?

   $(x^2 - y^2) / (x-y) = 7$

   $(x + y)^2 = 49$

   A. statement 1 alone is sufficient, but statement 2 alone is not sufficient to answer the question

   B. statement 2 alone is sufficient, but statement 1 alone is not sufficient to answer the question

   C. both statements taken together are sufficient to answer the question, but neither statement alone is sufficient

   D. each statement alone is sufficient

   E. statements 1 and 2 together are not sufficient, and additional data is needed to answer the question

9. Two socks are to be picked at random from a drawer containing only black and white socks. What is the probability that both are white?

The probability of the first sock being black is 1/3.

There are 24 white socks in the drawer.

A. statement 1 alone is sufficient, but statement 2 alone is not sufficient to answer the question

B. statement 2 alone is sufficient, but statement 1 alone is not sufficient to answer the question

C. both statements taken together are sufficient to answer the question, but neither statement alone is sufficient

D. each statement alone is sufficient

E. statements 1 and 2 together are not sufficient, and additional data is needed to answer the question

10. A bucket was placed under a dripping tap which was dripping at a uniform rate. At what time was the bucket full?

    The bucket was put in place at 2pm.

    The bucket was half full at 6pm and three-quarters full at 8pm.

    A. statement 1 alone is sufficient, but statement 2 alone is not sufficient to answer the question
    B. statement 2 alone is sufficient, but statement 1 alone is not sufficient to answer the question
    C. both statements taken together are sufficient to answer the question, but neither statement alone is sufficient
    D. each statement alone is sufficient
    E. statements 1 and 2 together are not sufficient, and additional data is needed to answer the question

11. It takes 3.5 hours for Mathew to row a distance of X km up the stream. Find his speed in still water.

(1) It takes him 2.5 hours to cover the distance of X km downstream.

(2) He can cover a distance of 84 km downstream in 6 hours.

A. Statement (1) ALONE is sufficient but Statement (2) ALONE is not sufficient.

B. Statement (2) ALONE is sufficient but Statement (1) ALONE is not sufficient.

C. BOTH Statements TOGETHER are sufficient, but NEITHER Statement alone is sufficient.

D. Each Statement ALONE is sufficient.

E. Statements (1) and (2) TOGETHER are NOT sufficient.

12. A man mixes two types of glues (X and Y) and sells the mixture of X and Y at the rate of $17 per kg. Find his profit percentage.

(1) The rate of X is $20 per kg.

(2) The rate of Y is $13 per kg.

A. Statement (1) ALONE is sufficient but Statement (2) ALONE is not sufficient.

B. Statement (2) ALONE is sufficient but Statement (1) ALONE is not sufficient.

C. BOTH Statements TOGETHER are sufficient, but NEITHER Statement alone is sufficient.

D. Each Statement ALONE is sufficient.

E. Statements (1) and (2) TOGETHER are NOT sufficient.

13.A bag contains 20 copper and 10 brass coins. If 9 of the coins are removed, how many copper coins are left in the box?

(1) Of the removed coins, the ratio of the number of copper coins to that of brass coins is 2 : 1

(2) Four of the first six coins removed are copper.

A. Statement (1) ALONE is sufficient but Statement (2) ALONE is not sufficient.

B. Statement (2) ALONE is sufficient but Statement (1) ALONE is not sufficient.

C. BOTH Statements TOGETHER are sufficient, but NEITHER Statement alone is sufficient.

D. Each Statement ALONE is sufficient.

E. Statements (1) and (2) TOGETHER are NOT sufficient.

14.Maria deposits $10,000 in a bank. What is the annual interest which the bank will pay to Maria?

(1) The interest must be paid once every six months.

(2) The rate of interest is 4%.

A. Statement (1) ALONE is sufficient but Statement (2) ALONE is not sufficient.

B. Statement (2) ALONE is sufficient but Statement (1) ALONE is not sufficient.

C. BOTH Statements TOGETHER are sufficient, but NEITHER Statement alone is sufficient.

D. Each Statement ALONE is sufficient.

E. Statements (1) and (2) TOGETHER are NOT sufficient.

15. Given that the length of the side of a square is 1 and that the length of the side is increased by x%. State whether the area of the square is increased by more than 10%.

(1) $x < 10$

(2) $x > 5$

A. Statement (1) ALONE is sufficient but Statement (2) ALONE is not sufficient.

B. Statement (2) ALONE is sufficient but Statement (1) ALONE is not sufficient.

C. BOTH Statements TOGETHER are sufficient, but NEITHER Statement alone is sufficient.

D. Each Statement ALONE is sufficient.

E. Statements (1) and (2) TOGETHER are NOT sufficient.

16. People in a club either speak French or Russian or both. Find the number of people in a club who speak only French.

(1) There are three hundred people in the club and the number of people who speak both French and Russian is 196.

(2) The number of people who speak only Russian is 58.

A. Statement (1) ALONE is sufficient but Statement (2) ALONE is not sufficient.

B. Statement (2) ALONE is sufficient but Statement (1) ALONE is not sufficient.

C. BOTH Statements TOGETHER are sufficient, but NEITHER Statement alone is sufficient.

D. Each Statement ALONE is sufficient.

E. Statements (1) and (2) TOGETHER are NOT sufficient.

17.Joe is older to Lloyd by five years. Ten years ago, John was 10 years older than Mary. What is Mary's age today?

(1) Mary's age today is three times the age of Joe.

(2) Lloyd today is 5 years old.

A. Statement (1) ALONE is sufficient but Statement (2) ALONE is not sufficient.

B. Statement (2) ALONE is sufficient but Statement (1) ALONE is not sufficient.

C. BOTH Statements TOGETHER are sufficient, but NEITHER Statement alone is sufficient.

D. Each Statement ALONE is sufficient.

E. Statements (1) and (2) TOGETHER are NOT sufficient.

18. A sum of $385 was divided among Jack, Pollock and Gibbs. Who received the minimum amount?

(1) Jack received 2/9 of what Pollock and Gibbs together received.

(2) Pollock received 3/11 of what Jack and Gibbs together received.

A. Statement (1) ALONE is sufficient but Statement (2) ALONE is not sufficient.

B. Statement (2) ALONE is sufficient but Statement (1) ALONE is not sufficient.

C. BOTH Statements TOGETHER are sufficient, but NEITHER Statement alone is sufficient.

D. Each Statement ALONE is sufficient.

E. Statements (1) and (2) TOGETHER are NOT sufficient.

19. 'n' is a natural number. State whether n (n?- 1) is divisible by 24.

(1) 3 divides 'n' completely without leaving any remainder.

(2) 'n' is odd.

A. Statement (1) ALONE is sufficient but Statement (2) ALONE is not sufficient.

B. Statement (2) ALONE is sufficient but Statement (1) ALONE is not sufficient.

C. BOTH Statements TOGETHER are sufficient, but NEITHER Statement alone is sufficient.

D. Each Statement ALONE is sufficient.

E. Statements (1) and (2) TOGETHER are NOT sufficient.

20. A policeman spots a thief and runs after him. When will the policeman be able to catch the thief?

(1) The speed of the policeman is twice as fast as that of the thief.

(2) The distance between the policeman and the thief is 400 meters.

A. Statement (1) ALONE is sufficient but Statement (2) ALONE is not sufficient.
B. Statement (2) ALONE is sufficient but Statement (1) ALONE is not sufficient.
C. BOTH Statements TOGETHER are sufficient, but NEITHER Statement alone is sufficient.
D. Each Statement ALONE is sufficient.
E. Statements (1) and (2) TOGETHER are NOT sufficient.

21. Who got the highest marks among Abdul, Baig and Chiman?

(1) Chiman got half as many marks as Abdul and Baig together got.

(2) Abdul got half as many marks as Baig and Chiman together got.

A. Statement (1) ALONE is sufficient but Statement (2) ALONE is not sufficient.
B. Statement (2) ALONE is sufficient but Statement (1) ALONE is not sufficient.
C. BOTH Statements TOGETHER are sufficient, but NEITHER Statement alone is sufficient.
D. Each Statement ALONE is sufficient.
E. Statements (1) and (2) TOGETHER are NOT sufficient.

22. Given that side AC of triangle ABC is 2. Find the length ofBC.

(1) AB is not equal to AC

(2) Angle B is 30 degrees.

A. Statement (1) ALONE is sufficient but Statement (2) ALONE is not sufficient.
B. Statement (2) ALONE is sufficient but Statement (1) ALONE is not sufficient.
C. BOTH Statements TOGETHER are sufficient, but NEITHER Statement alone is sufficient.
D. Each Statement ALONE is sufficient.
E. Statements (1) and (2) TOGETHER are NOT sufficient.

23. 50% of the people in a certain city have a Personal Computer and an Air-conditioner . What percent of people in the city have a personal computer but not an Air-conditioner.

(1) 60% of the people in the city have a Personal Computer.

(2) 70% of the people in the city have an Air-conditioner.

A. Statement (1) ALONE is sufficient but Statement (2) ALONE is not sufficient.

B. Statement (2) ALONE is sufficient but Statement (1) ALONE is not sufficient.

C. BOTH Statements TOGETHER are sufficient, but NEITHER Statement alone is sufficient.

D. Each Statement ALONE is sufficient.

E. Statements (1) and (2) TOGETHER are NOT sufficient.

24. Bags I, II and III together have ten balls. If each bag contains at least one ball, how many balls does each bag have?

(1) Bag I contains five balls more than box III.

(2) Bag II contains half as many balls as bag I.

A. Statement (1) ALONE is sufficient but Statement (2) ALONE is not sufficient.

B. Statement (2) ALONE is sufficient but Statement (1) ALONE is not sufficient.

C. BOTH Statements TOGETHER are sufficient, but NEITHER Statement alone is sufficient.

D. Each Statement ALONE is sufficient.

E. Statements (1) and (2) TOGETHER are NOT sufficient.

25. Given that (a + b)?= 1 and (a - b)?= 25, find the values 'a' and 'b'.

(1) Both 'a' and 'b' are integers.

(2) The value of 'a' = 2

A. Statement (1) ALONE is sufficient but Statement (2) ALONE is not sufficient.

B. Statement (2) ALONE is sufficient but Statement (1) ALONE is not sufficient.

C. BOTH Statements TOGETHER are sufficient, but NEITHER Statement alone is sufficient.

D. Each Statement ALONE is sufficient.

E. Statements (1) and (2) TOGETHER are NOT sufficient.

ANSWERS

1. E	2. D	3. B	4. C	5. E
6. D	7. E	8. A	9. C	10. B

11. Correct Answer: C

Solution:

Given that Mathew rows upstream with the speed of X / 3.5km/h.

Combining both the statements, we can calculate the downstream speed.

Downstream speed = 84 / 6 = 14 km/h.

Also, downstream speed = X / 2.5.

Or, X / 2.5 = 14.

Or X = 2.5 * 14 = 35 km.

Hence the upstream speed = X / 3.5

= 35 / 3.5

= 10 km/h.

So the speed in still water = (10 + 4) / 2 = 12 km/h.

Hence we need both the statements together to solve the question.

12. Correct Answer: E

Solution:

In order to find the profit or loss, the most important information we need to know is the ratio of X and Y. Neither of the statements provide us with any information regarding the ratios. Both the statements give only the rate of X and Y. Hence the given information is not sufficient to answer the given question.

13. Correct Answer: A

Solution:

Let the number of copper and brass coins removed be 2x and x respectively (from first statement). Now, given that $2x + x = 9$ or $x = 3$.

Hence we can conclude that the number of copper coins removed is 6. Hence statement 1 alone is sufficient.

Statement 2 gives only half the information. If the information about the other three coins (that were removed) had also been given, it would have been possible for us to find the answer. Hence statement 2 alone is insufficient.

14.Correct Answer:B

Solution:

The only thing we need to know while calculating the annual interest is the rate of interest which is given in only statement 2. Hence it is the only statement which is sufficient.

15.Correct Answer: B

Solution:

According to the first statement, if $x < 10$, it can be any number between 0 to 9. In such case, the area may or may not increase by more than 10%. Hence statement 1 alone is insufficient.

Now, if we take x=" 5," then the area increases by 10.25%. Hence for every value of $x > 5$, the area has to increase by more than 10%. Hence statement 2 alone is sufficient.

16.Correct Answer: C

Solution:

Let the number of people who speak French be p, that of who speak both be q, and that of who speak only Russian be r.

According to statement 1, $p + q + r = 300$ and $q = 196$

But we need the value of r to calculate the value of p. Hence statement 1 alone is insufficient.

Statement 2 gives only the value of r. Hence p and q cannot be found by statement 2 alone. Hence statement 2 alone is also insufficient.

However, from both the statements, we get, $p = 300 - q - r = 300 - 196 - 58 = 46$. Thus the value can be found. Hence we need both the statements together to answer the given question.

17. Correct Answer: C

Solution:

Let the ages of Joe, Lloyd, John and Mary be p, q, r and s.

It is also given that $p = q + 5$, and $r = s + 10$. Now, we have to find the value of s.

According to statement 1, $s = 3p$. However as their present ages are not given, hence 1 alone is insufficient to find the answer.

According to statement 2, $q = 5$. Hence the value of p can be calculated to be 10. However as statement 2 does not alone give any relation between p and s, it alone is insufficient to answer the question.

Using both the statements together, we find that $s = 3p = 3 * 10 = 30$. Hence we require both the statements together to answer the given question.

18. Correct Answer: C

Solution:

Let the amount received by Jack, Pollock and Gibbs be x, y and z respectively.

Also, $x + y + z = 385$. ------- (1)

According to statement 1, $x = (2/9)(y + z)$. -------- (2)

This gives us two equations. But there are three unknowns to be found. Hence statement 1 alone is insufficient.

According to statement 2, $y = (3/11)(x + z)$ ------- (3)

And $x + y + z = 385$ [equation ---- (1)]

Again we have two equations and three unknowns. Hence statement 2 alone is also insufficient to find the answer. However, if we combine both the statements together, we get three different equations and three unknowns. Hence we need both the statements to find the answer.

19. Correct Answer: B

Solution:

According to statement 1, n is a multiple of 3.

Now, say if we take n = 3, the expression is divisible, but in case, we put n = 6 or 12, then the expression is not divisible by 24. Hence statement 1 alone is insufficient.

Statement 2 alone states that n is odd. Now, if we put any odd value in place of n, we find that the expression is divisible by 24. Hence option 2 alone is sufficient.

20. Correct Answer: E

Solution:

Statement 1 only gives the speeds of both, the thief and the policeman, which cannot be helpful in finding the time. Hence statement 1 alone is insufficient.

Similarly statement 2 gives no clue about the time, it only gives the distance between the two. Hence it alone is also insufficient.

Combining both the statements would also not help us knowing the time. Hence the answer cannot be found from the given information.

21. Correct Answer: C

Solution:

Let the marks of Abdul, Baig and Chiman be X, Y and Z respectively.

According to statement 1, Z = ?(X + Y) or 2Z = X + Y ------- (1)

We have only one equation but two unknowns. Hence statement 1 alone is insufficient.

According to statement 2, X = ?(Y + Z) or 2X = Y + Z ------- (2)

Again, we have only one equation but two unknowns. Hence statement 2 alone is also insufficient.

However, if we combine both the statements, we get two different equations from which we can find the answer.

22. Correct Answer: E

Solution:

The given properties of the triangle are insufficient to provide any relationship between the sides or the angles of the triangle. It is given that angle Q is 30 degree and side PR is equal to 2. QR could be of any length, which cannot be deduced from the given information.

23. Correct Answer: A

Solution:

According to statement 1, 60 - 50 = 10% of people have a Personal Computer but not an Air-conditioner. Hence statement 1 alone is sufficient to answer the given question.

Statement 2 only helps in finding out what percentage of people have Air-conditioner and not the percentage of people having Personal computer. Hence it is insufficient to derive the answer.

24. Correct Answer: C

Solution:

From statement 1, only two combinations are possible. Bag III contains 1 and bag I contains 6 or bag III contains 2 and bag I contains 7 balls. This information alone is insufficient to answer the given Question.

From statement 2, there are three possibilities; bag II has 1, bag I has 2; bag II has 2, bag I has 4, and bag II has 3, bag I has 6 balls. Hence it also is insufficient.

If both the statements are combined, we get the possible answer, bag I has 6, bag III has 1 and bag II has 3 balls. Hence we need both the statements together to answer the given question.

25. Correct Answer: B

Solution:

On solving both the equations given in the main question, we get ab = - 6. ------- (1)

Now according to statement 1, a and b are integers, they can be [2, - 3]; [- 2, 3]; [1, - 6]; [6, - 1], etc. So statement 1 alone is insufficient.

According to statement 2, a = 2. Hence b = - 2 ------ [from equation ----(1)]

Hence statement 2 alone is sufficient.Math Concepts You Should Know

IV. Numerical Reasoning

在數字推理題型中，每道試題中呈現一組按某種規律排列的數字，但這一數列中有意地空缺了一項，要求考生仔細觀察這一組數列，找出數列的排列規律，從而根據規律推導出空缺項應填的數字，然後用戶答題區提供的四個選項中選出你認為最合理、最適合的選項。

首先找出相鄰兩個（特別是第一、第二個）數字間的關係，迅速將這種關係推到下一個數字相鄰間的關係，若得到驗證，説明找到了規律，就可以直接推出答案；若被否定，馬上改變思考方向和角度，提出另一種數量關係假設。如此反復，直到找到規律為止。有時也可以從後面往前推，或者「中間開花」向兩邊推，都可能是較為有效的。解答此類試題的關鍵是找出數位排列時所依據的某種規律，通過相鄰兩數位間關係的兩兩比較就會很快的找到共同特徵，即規律。規律被找出來，答案自然就出來了。

在進行此項測驗時要善於總結經驗前應加強練習，了解有關出題形式，考試時就能得心應手。當然，在推導數量關係時，必然會涉及到許多計算，但你儘量不用筆算或少用筆算，而多用心算，這樣可以縮短做題時間，用更多的時間做其他題目。

Numerical Reasoning
數字推理題的題型

1. 等差數列及其變式

例題： 1, 4, 7, 10, 13, ( )

A.14　　B.15　　C.16　　D.17

答案為C。我們很容易從中發現相鄰兩個數字之間的差是一個常數3，所以括弧中的數字應為16。等差數列是數位推理測驗中排列數位的常見規律之一。

例題： 3, 4, 6, 9, ( ), 18

A.11　　B.12　　C.13　　D.14

答案為C。仔細觀察，本題中的相鄰兩項之差構成一個等差數列1, 2, 3, 4, 5……，因此很快可以推算出括弧內的數位應為13，像這種相鄰項之差雖不是一個常數，但有著明顯的規律性，可以把它看作等差數列的變式。

2.「兩項之和等於第三項」型

例題： 34, 35, 69, 104, ( )

A.138　B.139　C.173　D.179

答案為C。觀察數字的前三項，發現第一項與第二項相加等於第三項，34+35=69，在把這假設在下一數字中檢驗，35+69=104，得到驗證，因此類推，得出答案為173。前幾項或後幾項的和等於後一項是數字排列的又一重要規律。

3. 等比數列及其變式

例題： 3, 9, 27, 81, ( )

A.243　B.342　C.433　D.135

答案為A。這是最一種基本的排列方式，等比數列。其特點為相鄰兩項數字之間的商是一個常數。

例題： 8, 8, 12, 24, 60, ( )

A.90　B.120　C.180　D.240

答案為C。雖然此題中相鄰項的商並不是一個常數，但它們是按照一定規律排列的：1，1.5，2，2.5，3，因此答案應為60×3=180，像這種題可視作等比數列的變式。

4. 平方型及其變式

例題： 1, 4, 9, ( ), 25, 36

A.10　B.14　C.20　D.16

答案為D。這道試題考生一眼就可以看出第一項是1的平方，第二項是2的平方，如此類推，得出第四項為4的平方16。對於這種題，考生應熟練掌握一些數字的平方得數。如：

10的平方=100
11的平方=121
12的平方=144
13的平方=169
14的平方=196
15的平方=225

例題： 66, 83, 102, 123, ( )

A.144　B.145　C.146　D.147

答案為C。這是一道平方型數列的變式，其規律是8，9，10，11的平方後再加2，因此空格內應為12的平方加2，得146。這種在平方數列的基礎上加減乘除一個常數或有規律的數列，可以被看作是平方型數列的變式，考生只要把握了平方規律，問題就可以化繁為簡了。

5. 立方型及其變式

例題： 1, 8, 27, ( )

A.36　B.64　C.72　D.81

答案為B。解題方法如平方型。我們重點説説其變式

例題： 0, 6, 24, 60, 120, ( )

A.186　B.210　C.220　D.226

答案為B。這是一道比較有難道的題目。如果你能想到它是立方型的變式，就找到了問題的突破口。這道題的規律是第一項為1的立方減1，第二項為2的立方減2，第三項為3的立方減3，依此類推，空格處應為6的立方減6，即210。

6. 雙重數列

例題： 257, 178, 259, 173, 261, 168, 263, ( )

A.275　B.178　C.164　D.163

答案為D。通過觀察，我們發現，奇數項數值均為大數，而偶數項都是小數。可以判斷，這是兩列數列交替排列在一起而形成的一種排列方式。在這類題目中，規律不能在鄰項中尋找，而必須在隔項中尋找，我們可以看到，奇數項是一個等差數列，偶數項也是一個等差數列，因此不難發現空格處即偶數項的第四項，應為163。也有一些題目中的兩個數列是按不同的規律排列的，考生如果能判斷出這是多組數列交替排列在一起的數列，就找到了解題的關鍵。

給你一個數列，但其中缺少一項，要求你仔細觀察數列的排列規律，然後從四個選項中選出你認為最合理的一項來填補空缺項。

1. 6, 24, 60, 120, ( )

A. 186 B. 200

C. 210 D. 220

2. 345, 268, 349, 264, 354, 259, 360, ( )

A. 366 B. 255

C. 370 D. 253

3. 2, 9, 16, 23, 30, ( )

A. 35 B. 37

C. 39 D. 41

4. 1, 3, 5, 7, 9, ( )

A. 7 B. 8

C. 11 D. 12

5. 1, 8, 27, ( )

A. 64 B. 72

C. 81 D. 36

6. 1, 4, 9, 16, ( ) ,36

A. 23　B. 25

C. 27　D. 31

7. 2, 5, 8, 11, ( )

A. 12　B. 13

C. 14　D. 15

8. 1, 5, 6, 11, 17, ( )

A. 24　B. 28

C. 31　D. 33

9. 6, 10, 18, 34, ( )

A. 64　B. 66

C. 68　D. 70

10. 3, 4, 6, 9, ( ) , 18

A. 11　B. 12

C. 13　D. 15

11. 345, 268, 349, 264, 353, 260, 357, ( )

A. 36　　B. 255

C. 370　　D. 256

12. 1, 5, 6, 11, 17, ( )

A. 24　　B. 28

C. 31　　D. 33

13. 118, 199, 226, 235, ( )

A. 238　　B. 246

C. 253　　D. 255

14.

4　5
56
10　9

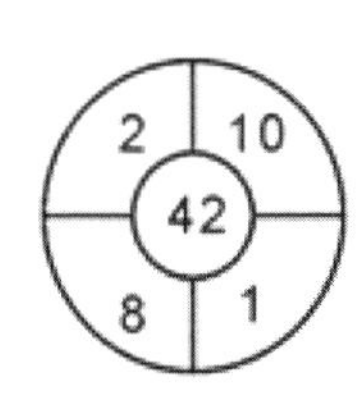

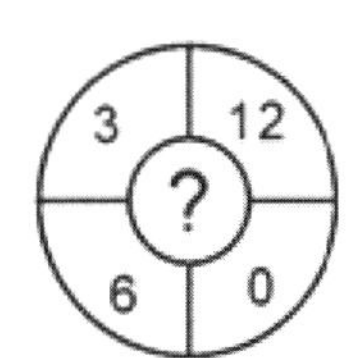

A. 21　B. 42　C. 36　D. 57

15.

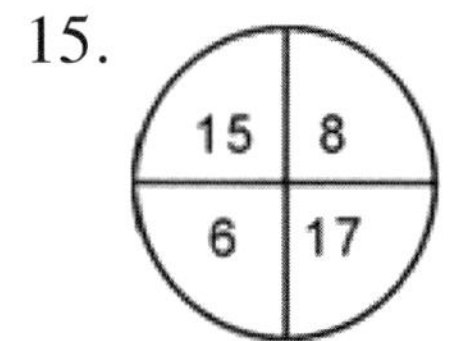

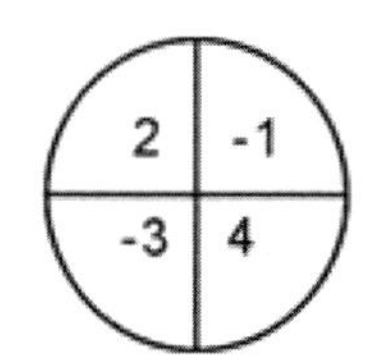

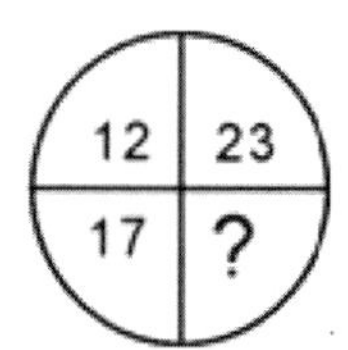

A. 16　B. 18　C. 20　D. 22

16.

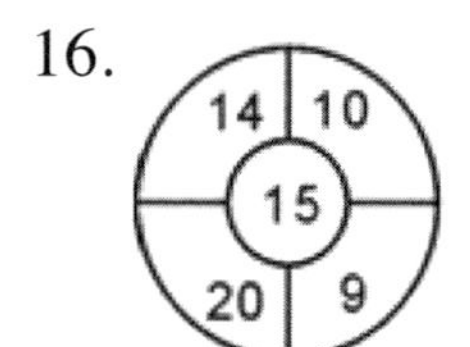

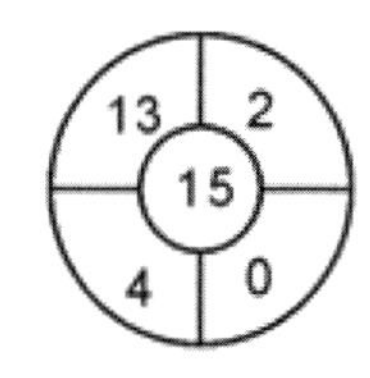

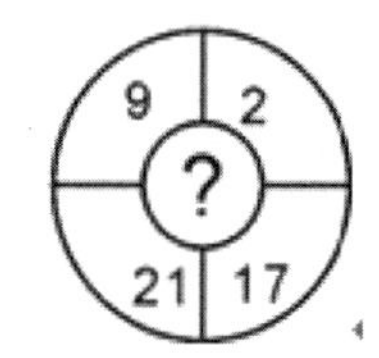

A. 15　B. 14　C. 11　D. 9

17.

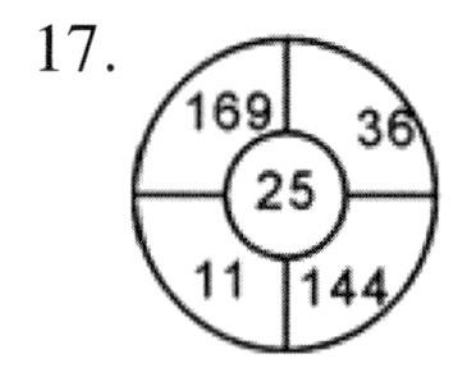

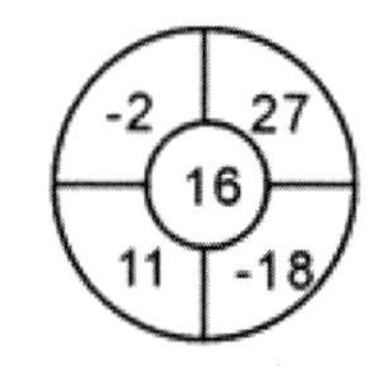

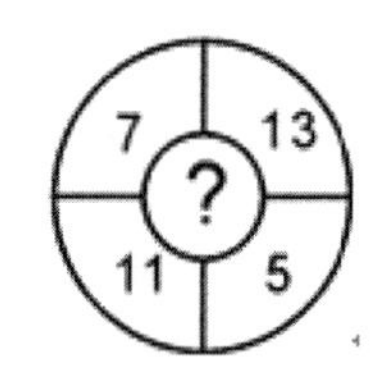

A. 9　B. 6　C. 4　D. 2

18.

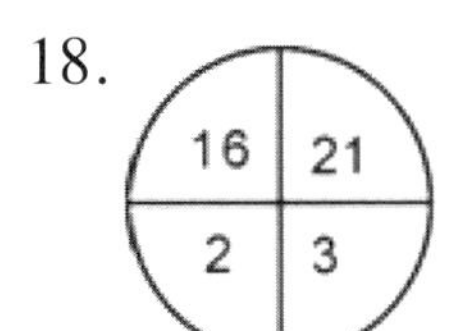

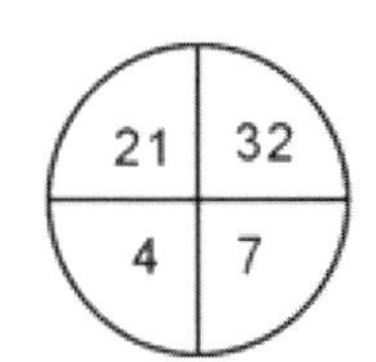

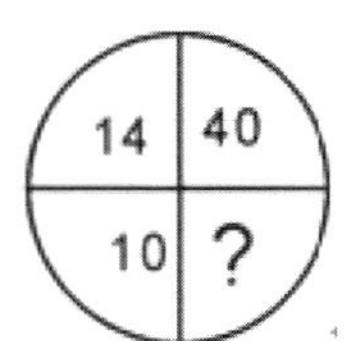

A. 17　B. 16　C. 13　D. 19

19.

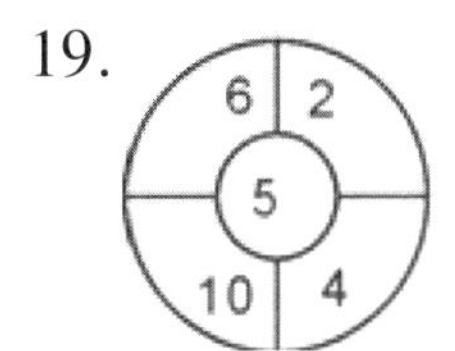

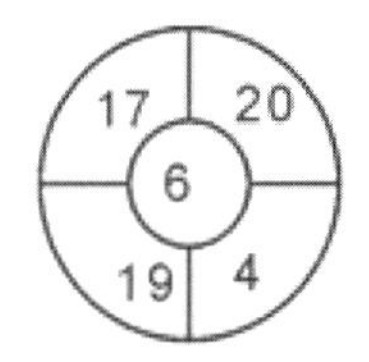

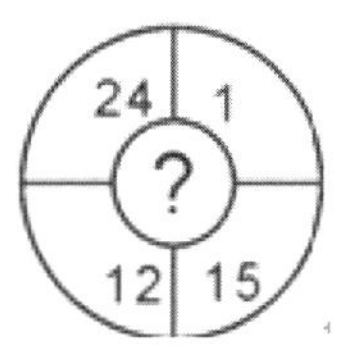

A. 7　B. 8　C. 10　D. 13

20.

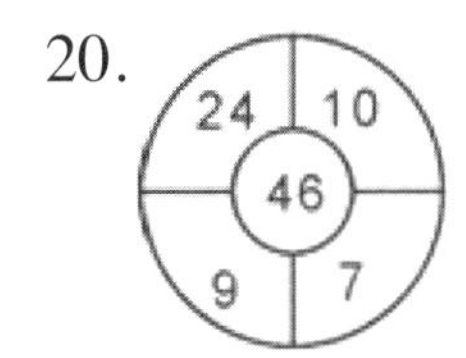

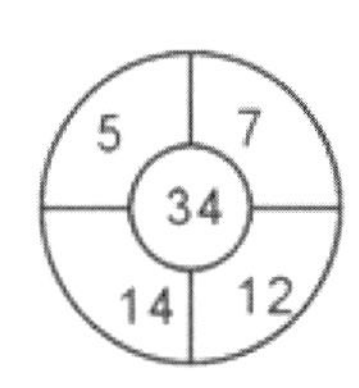

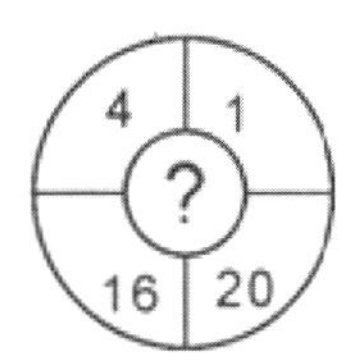

A. 37　B. 29　C. 27　D. 24

答案與解釋

1. 答案：n的立方減去n，答案為C。

2. 答案：奇數位置上的數按4、5、6遞增，偶數位置上的數按4、5、6遞減，答案為D。

3. 答案：這一數列的排列規律是前一個數加7等於後一個數，故空缺項應為37。正確答案為B。

4. 答案：這是一個奇數數列，成等差方式排列的，每相鄰兩數字均相差2，所以括弧中的數字應是11，即選項C為正確答案。以等差數列的方式排列數位，是數位推理測驗排列數位的規律之一，也是一種很簡單的排列方式。

5. 答案：這是各項分別為1，2，3，4的立方的數列。答案為A。

6. 答案：連續自然數的平方，答案為B。

7. 答案：前一個數加3等於後一個數，答案為C。

8. 答案：前面相鄰兩數的和等於下一個數，答案為B。

9. 答案：前一個數的兩倍減去2等於下一個數，答案為B。

10. 答案：相鄰兩數間的差值成等差數列，答案為C。

11. 答案：仔細觀察可以發現，隔項之間分別為遞增的等差數列345，349，353，357和遞減的等差數列268，264，260，( )。顯然可推知答案為D。

12. 答案：第一個數位1與第二個數位5之和正好是第三個數字6，而第二個數位5與第三個數位6之和正好是第四個數字11。繼續往下推，第三個數位6與第四個數位11之和正好是第五個數字17。因此，括弧中的數位應該是第四個數位11和第五個數位17的和，即28，故選項B為正確答案。

13. 答案：這題目並不直接表現為等比數列，但是我們可以經過簡化、處理，得到一個等比數列，將題中後項與前項相依相減，得到81，27,9，( )的等比數列，可知( )應為3。由此可推知答案為A。

14. 答案：B。該數列的規律是中間的數字為其他四個數字之和的兩倍，故問號處應為$2\times(3+12+6+0)=42$。

15. 答案：B。上邊兩數之和等於下邊兩數之和，故問號處應為12+23-17＝18。

16. 答案：C。該數列的規律是左邊兩數之和減去右邊兩數之和為中間數字，故問號處應為(21+9)-(2+17)＝11。

17. 答案：D。對角上的兩數之差為中間數字，故問號處應為7-5＝13-11＝2。

18. 答案：B。該數列的規律是16+3＝21-2，21+7＝32-4，故問號處應為40-10-14＝16。

19. 答案：C。中間數字為左邊兩數之和減去右邊兩數之和的差的一半。故問處應為(24+12-1-15)×　＝10。

20. 答案：A。中間數字為其他四個數字之和再減去4。故問號處應為(4+1+16+20)-4＝37。

V. Interpretation of Tables and Graphs

This is a test on reading and interpretation of data presented in tables and graphs. You are required to find answers based on the information provided in the question.

Study the graphs and tables to answer the questions.

Graph 1 (Question 1 to 3)

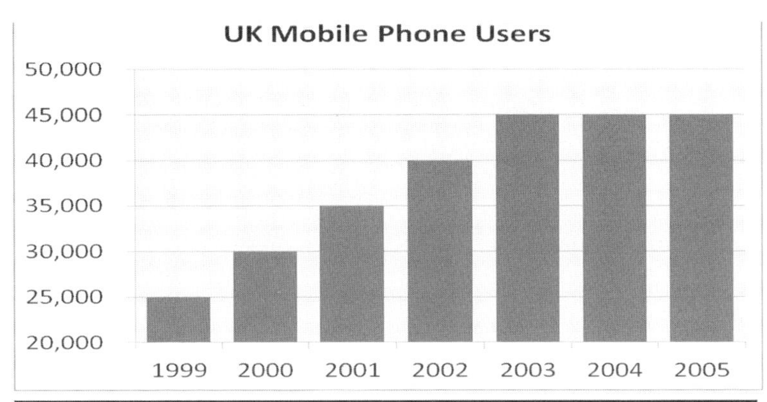

Market Share in 2004						
Nokia	Motorola	Samsung	Siemens	LG	Ericson	Others
30%	15%	13%	7%	7%	6%	22%

1. How many mobile phone users were using Samsung handsets in 2004?

   A. 5,850 B. 8,775 C. 2,165 D. 625

2. If Nokia's market share was 32% in 2002, how does the number of Nokia users in 2002 compare with that of 2004?

   A. 700 fewer B. 700 more

   C. 300 fewer D. 300 more

3. In 2004, how many more mobile phone users would LG require to equal that of Motorola?

   A. 3,600 B. 1,350 C. 1,900 D. 500

Graph 2 (Question 4 to 6)

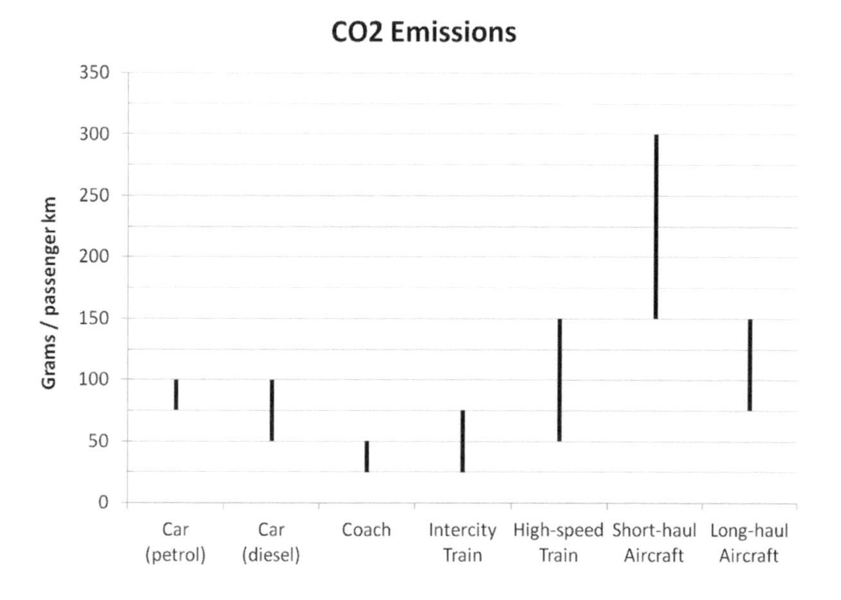

4. Approximately, what is the maximum amount of CO_2 that could be produced from one passenger travelling 200km by Intercity Train?

   A. 15kg B. 1.5kg

   C. 20kg D. 2kg

5. What is the difference between minimum and maximum CO_2 emissions for one passenger travelling 50km by Coach, 100km by High-speed Train, then 20km by Intercity Train?

   A. 12,250 B. 17,250

   C. 8,850 D. 20,465

6. 2,000 passengers pass through a short-haul airport in one day. If each passenger travels an average of 400km, what are the minimum CO_2 emissions from the airport?

   A. 1.2 million kg/day B. 120,000 kg/day

   C. 120 kg/day D. 12 kg/day

Graph 3 (Question 7 to 9)

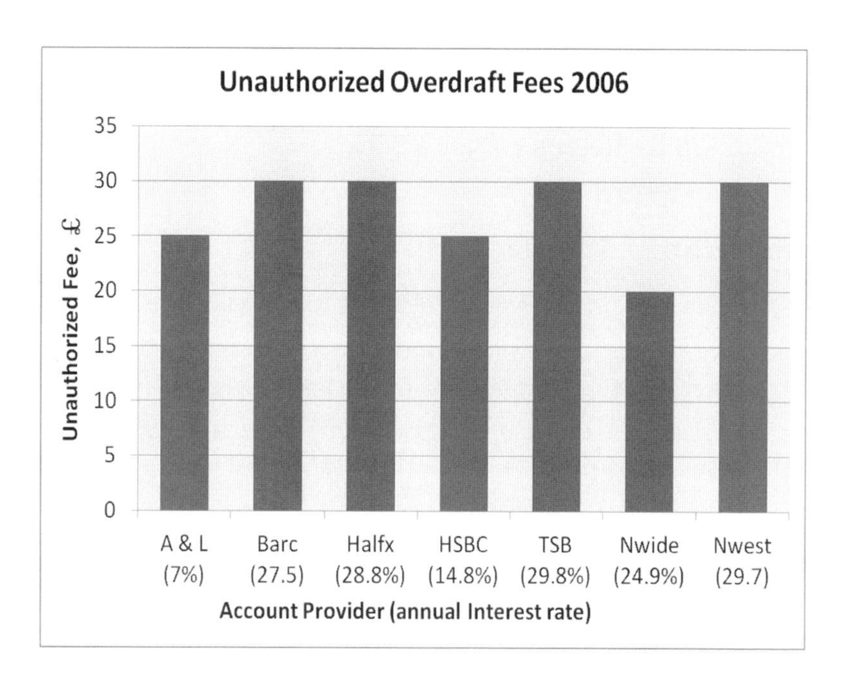

7. Using simple interest, which account provider would charge the most for an unauthorized overdraft of £100 over the period of one year?

   A. Barc B. Halfx C. TSB D. NWest

8. In 2006, A&L charged 8,000 of its customers for an unauthorized overdraft, and Nwide charged 8,500 of its customers for an unauthorized overdraft. What is the difference between the two banks in Unauthorized Fees, ignoring any interest charges?

   A. £30,000 B. £2,500 C. £40,000 D. £12,500

9. Next year, Halfx is decreasing its Unauthorized Fee by 6%. How much will this new fee be?

   A. £28.2 B. £36.1 C. £28.0 D. £26.4

Graph 4 (Question 10 to 11)

The graph shows the values of Stock Z for the year 1987.

10. The greatest increase for Stock Z, compared to the month before, occurred during which month?

    A. April B. May C. December

    D. March E. August

11. For the overall year, 1987, what was the percent increase of Stock Z?

    A. 250% B. 100% C. 50%

    D. 20% E. 0%

Graph 5 (Question 12)

Of all the students sent to the principal's office on one school day, each was asked his/her soda consumption for that day. The graph above represents each child's answer.

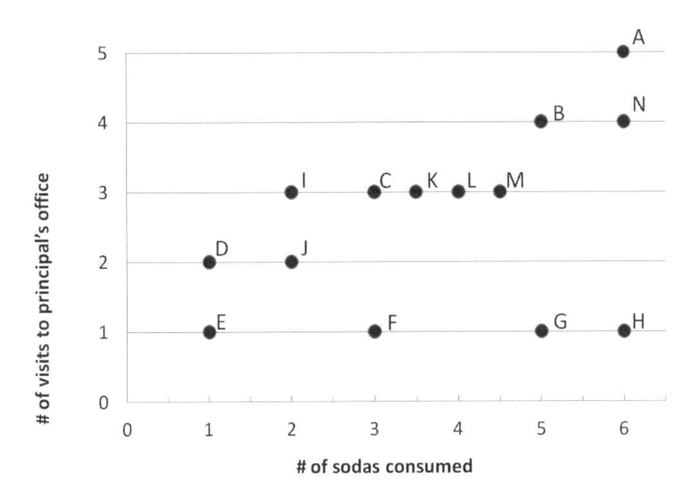

12. If you were to draw a line showing the average ratio of principal's office visits to sodas consumed, this line would most likely pass through point

A. K B. I C. H

D. G E. F

Graph 6 (Question 13 to 14)

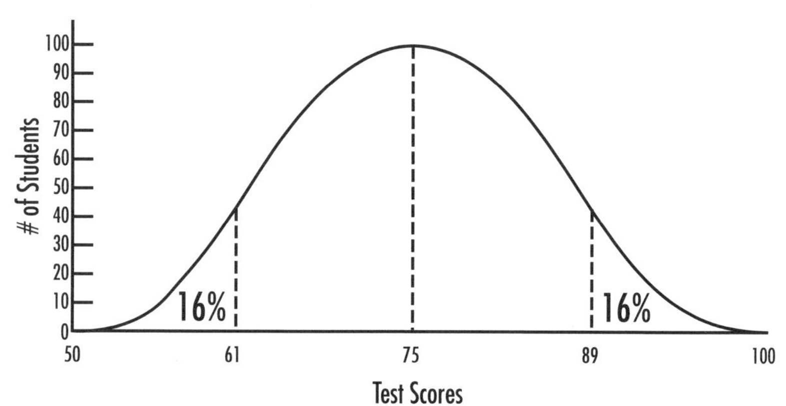

13. What percent of the students scored between 61 and 89?

    A. 2% B. 14% C. 50%
    D. 68% E. 74%

14. What are the mean, median, and mode, respectively, of the test scores?

    A. 89, 80, 75 B. 75, 75, 75
    C. 61, 75, 89 D. 89, 75, 61
    E. 75, 61, 89

Graph 7 (Question 15 to 16)

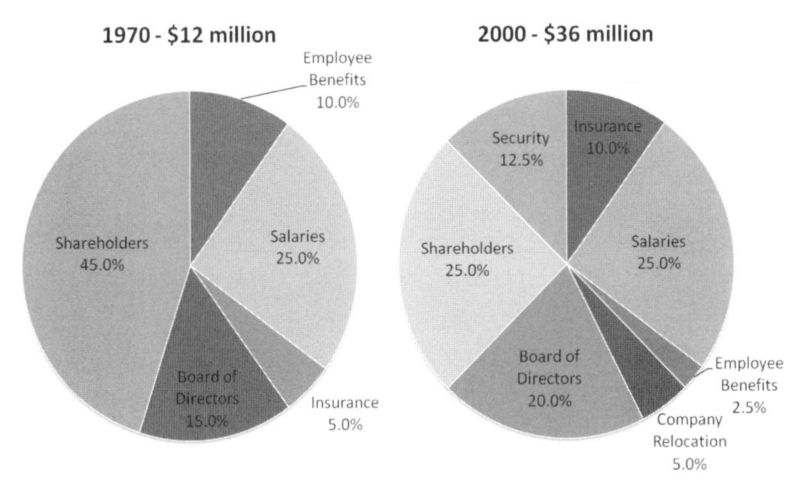

15. In 2000, Company X had to add security costs and company relocation to its profit distribution. These expenses were taken out of employee benefits. What was the approximate percent decrease in employee benefits from 1970 to 2000?

A. 7.5% B. 17.5% C. 25% D. 33.3% E. 35%

16. If the Board of Directors consists of the same eight people for both 1970 and 2000, then between 1970 and 2000, each board member, on average, saw his or her portion of profits increase by how much?

A. $225,000 B. $675,000

C. $900,000 D. $1,200,000

E. $1,800,000

Graph 8 (Question 17 to 19)

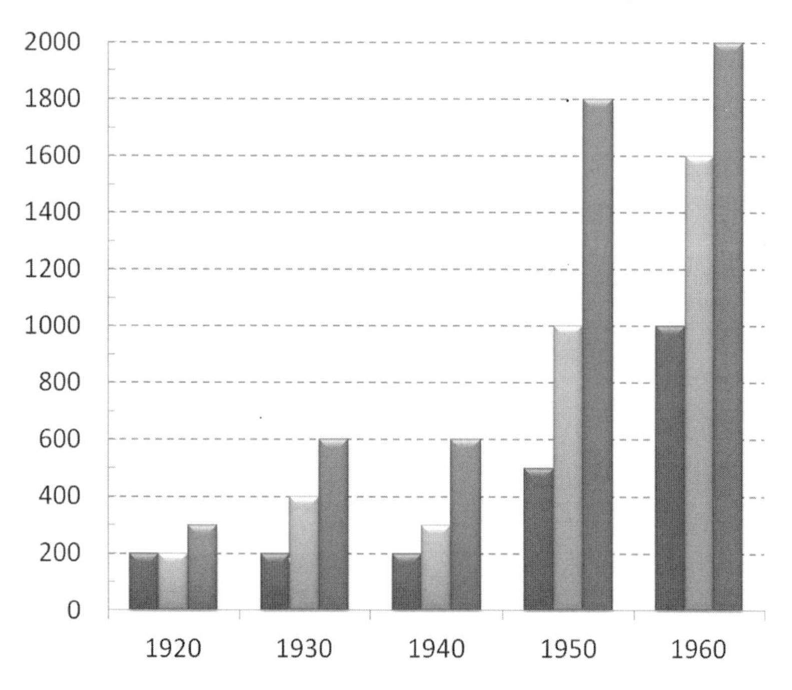

17. During the one decade that Red River's population decreased, by approximately what percent did the town's total population change?

A. 25% B. 20% C. 15% D. 10% E. 5%

18. During which decade did Big City experience the greatest percentage in population growth?

A. 1920–1930 B. 1930–1940 C. 1940–1950
D. 1950–1960 E. 1960–1970

19. During which decade did two of the towns have the same population?

A. 1920 B. 1930 C. 1940 D. 1950 E. 1960

Graph 9 (Question 20 to 21)

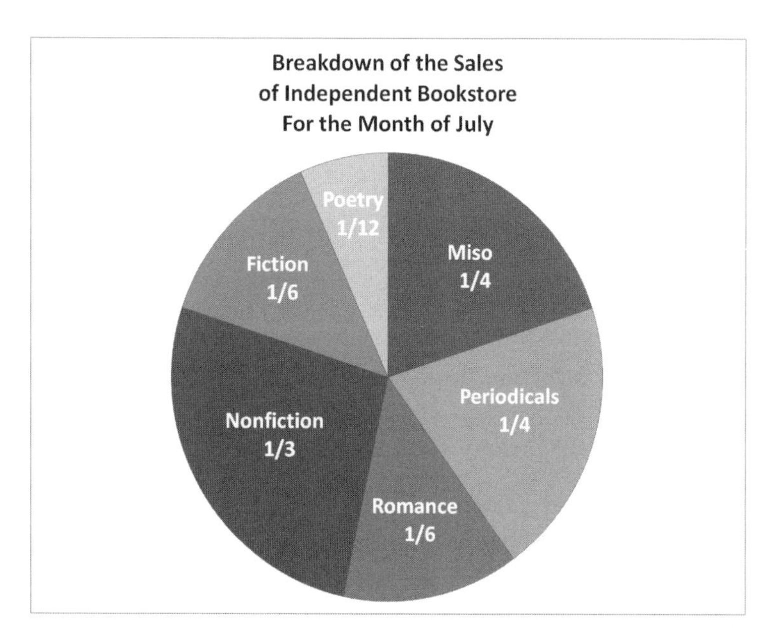

20. If the total revenue for the month of July was $24,000, then how much did the store make in sales on books about poetry?

A. $1,200　B. $2,000　C. $4,000
D. $6,000　E. $12,000

21. If the categories of romance and fiction were put together, then their total in sales would be equal to which other category of book?

A. Periodicals　B. Poetry　C. Nonfiction
D. Misc.　E. None of the above

Graph 10 (Question 22 to 28)

Use the information provided in the two pie charts provided below. The total contribution to the GDP by the seven sectors mentioned in the pie charts in the year 1999 was Rs.289640 crores and Rs.317000 crores in the year 2000.

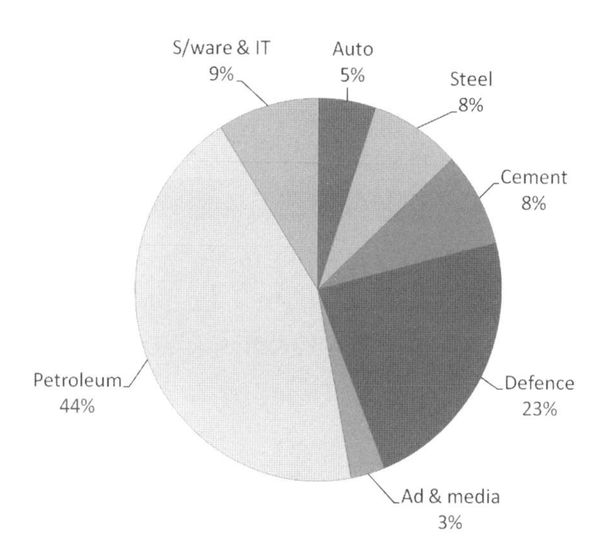

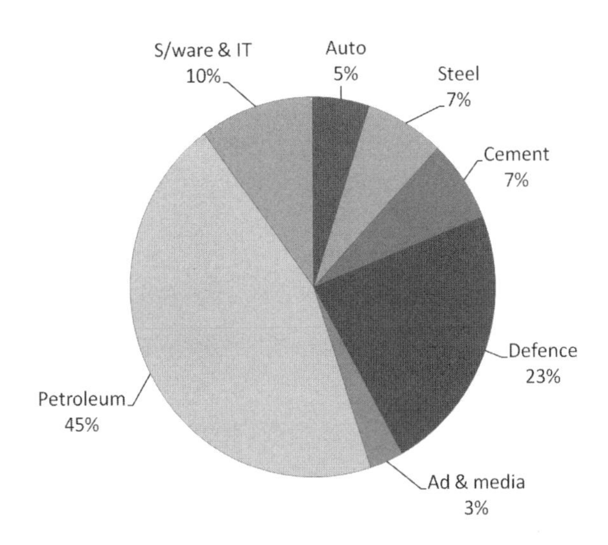

22. Which of the industry sectors witnessed the maximum rate of growth during the period 1999-2000?

A. Petroleum　　B. Software & IT

C. Ad & media　　D. Cement

23. Which of the industry sectors witnessed a negative growth during the period 1999-2000?

A. Auto　B. Defence　C. Steel　D. Petroleum

24. What was the rate of growth witnessed by the Software & IT sector during this period?

A. 1%　B. 12%　C. 33%　D. 22%

25. What was the rate of growth witnessed by the Petroleum sector during this period?

A. 1.1% B. 12% C. 7.5% D. -8%

26. What was the rate of growth shown by the non-petroleum sectors between 1999-2000?

A. -4% B. 4% C. 7% D. 12%

27. Between 1999 and 2000 which other industry witnessed a growth rate similar to that of the defence sector?

A. Ad & media
B. Auto
C. Software & IT
D. (A) & (B)

28. The amount contributed by Software & IT sector in 1999 was 180% of the amount contributed by

A. Steel in 1999
B. Auto in 1999
C. Ad & media in 2000
D. Defence in 1999

Graph 11 (Question 29 to 33)

BUS SCHEDULE - Table

Saturday AM Westbound Bus Schedule (Title)

Bus Departure Time (Label)				
7:16	7:28	7:32	7:36	8:10
8:01	8:13	8:17	8:21	8:55
9:31	9:43	9:47	9:51	10:28
10:16	10:28	10:32	10:36	11:13
11:01	11:13	11:17	11:21	11:58
12:31	12:43	12:47	12:51	1:28
Great Mill	May-Rich	May-Green	May-Warren	Center Square

Stations Served (Label)

29. When is the time of the last bus leaving May-Green that will get you to Center Square by 9:00 a.m.?

A. 8:01 a.m. B. 8:17 a.m.

C. 7:16 a.m. D. 8:21 a.m.

30. How many minutes does it take the bus to get from May-Rich to May-Warren?

A. 6 minutes B. 8 minutes

C. 20 minutes D. 4 minutes

31. There are six rows and five columns of information in this schedule. How many cells of information are in the table?

A. 2 B. 5 C. 6 D. 30

32. What time does the bus leave Great Mill that gets to Center Square at 10:28 a.m.?

A. 9:31 a.m. B. 9:43 a.m.
C. 8:55 a.m. D. 8:01 a.m.

33. If you want to know at what time the 9:31 a.m. Great Mill bus will arrive at your stop May-Warren you can just look at

A. column four B. row four, column three
C. column three D. row three, column four

Graph 12 (Question 34 to 36)

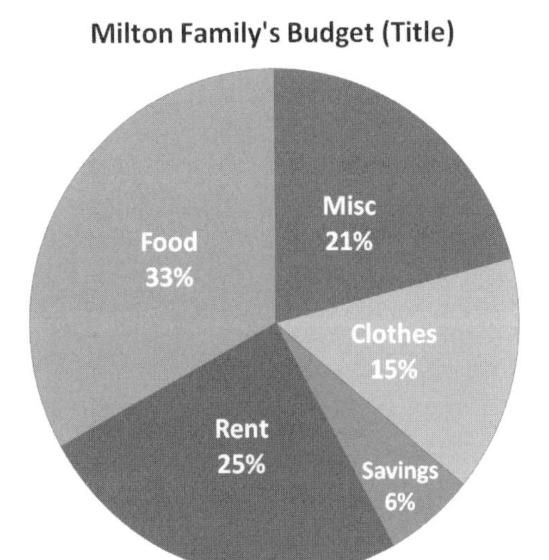

34. Which of the following statements is FALSE?

A. The family spends less on clothes than food.

B. The family spends more on rent than clothes and savings.

C. The family spends 100% of their income and saves nothing.

D. The family spends more than 50% of their income on food and rent.

35. What conclusion can be drawn from the Milton Family Budget pie graph?

A. The most money is spent on rent.

B. As income increases the amount of money saved will stay constant.

C. $3,300 a year will feed the Milton family.

D. If the family income decreases, less money will be saved.

36. If the Milton family moved in with another family and only had to spend half as much of their income on rent, they would pay what %?

A. 25% B. 50% C. 12.5% D. 0%

Graph 13 (Question 37 to 42)

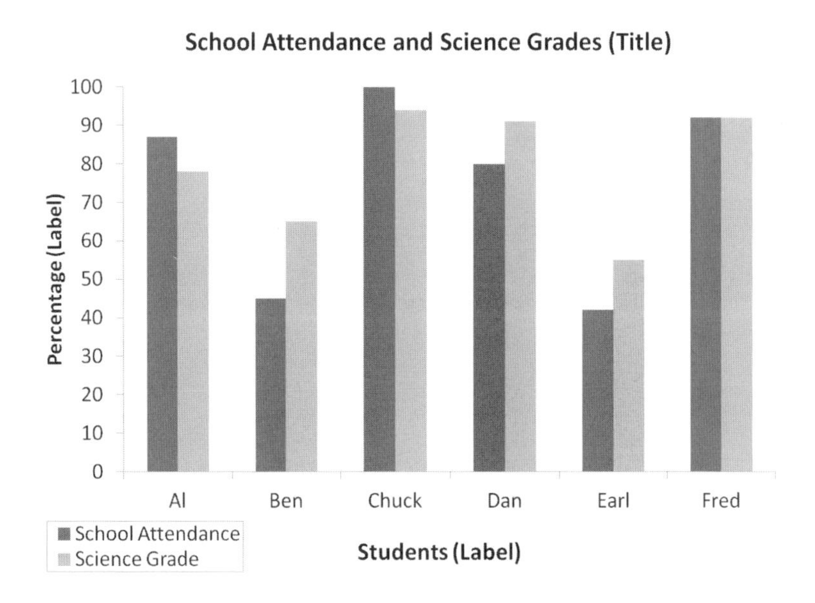

37. If a science grade below 55% is failing, how many students failed science?

A. 1　B. 0　C. 2　D. 4

38. The student with the best grade and attendance was

A. Al　B. Chuck　C. Dan　D. Fred

39. The student with the same school attendance and grade percentage was

A. Fred　B. Dan　C. Chuck　D. Ben

40. The bar graph shows the correlation of school attendance and the student's science grade. Study the bar graph to see which statement is FALSE.

A. Ben and Earl's attendance at school and their grades are poor.

B. Three students had higher grade percentages than the percentage of school attendance.

C. If Ben and Earl had worked together they could have gotten 100% in science.

D. The higher percentage of grades, the higher the percentage of attendance.

41. Ben was what number in order from the student with the top grade to the lowest grade of the six listed?

A. 2 B. 3 C. 6 D. 5

42. Which statement of the following is true?

A. Ben and Earl will not be promoted to the next grade.

B. School attendance affects the science grade.

C. Dan and Al attended school about the same number of days.

D. Chuck would get a better grade if he did not attend school so regularly.

Graph 14 (Question 43 to 45)

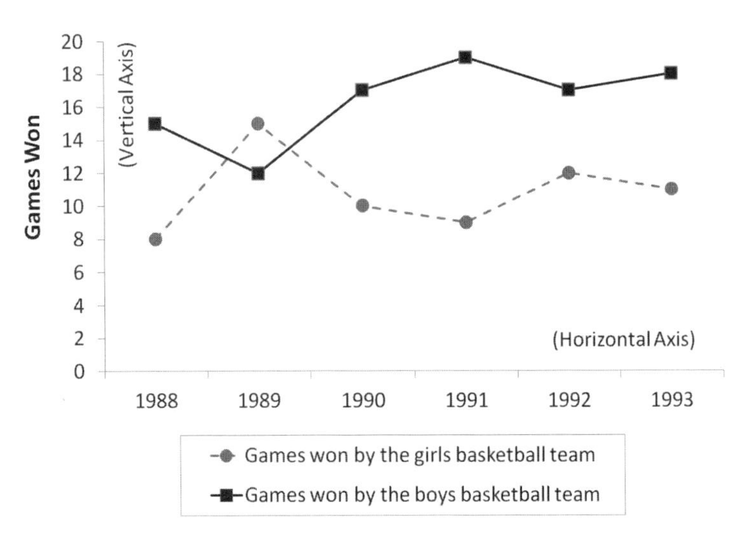

43. Looking at the line graph, in which year did the boys win twice as many games as the girls?

A. 1993 B. 1990

C. 1988 D. 1991

44. In what year did the girls win more games than the boys?

A. 1989 B. 1990

C. 1991 D. 1992

45. Which statement of the following is true?

A. There are more boys than girls out for basketball.

B. In each year that the boys won fewer games than the previous year, the girls won more games than the previous year.

C. The boys increased the number of games won each year from 1989-1993.

D. Overall the girls tend to win more basketball games.

ANSWERS

1. A: 13% of 45,000 is 5,850.

2. A : In 2002 the number of Nokia handsets was 32% of 40,000. In 2004: 30% of 45,000.

3. A: In 2004 LG had 7% of 45,000 whilst Motorola had 15%.

4. A: Maximum for Intercity train is 75g/passenger km. In 200km this is 15,000g.

5. A: The minimum emissions for coach is 25 g/passenger km, 50 for high speed train and 25 for intercity train. Work out the minimum and maximum for each leg.

6. B: 2,000 passengers x 400km x 150g/passenger km = 120,000kg.

7. C: The figures in brackets give the annual percentage rate. Multiply this by £100 and add the unauthorized overdraft fee (y-axis).

8. A: £25 x 8000 for A&L and £20 x 8500 for Nwide.

9. A: £30 x 0.94 is 28.2.

10. D	11. D	12. A	13. D	14. B	15. C
16. B	17. A	18. C	19. A	20. B	21. C
22. B	23. C	24. D	25. B	26. C	27. A
28. B	29. B	30. B	31. D	32. A	33. D
34. C	35. D	36. C	37. A	38. B	39. A
40. C	41. D	42. B	43. D	44. A	45. B

PART THREE

模擬試題測驗一

能力傾向測試
模擬測驗(一)

限時四十五分鐘

I. 演繹推理(8題)

請根據以下短文的內容，選出一個或一組推論。請假定短文的內容都是正確的。

1. 作家在其晚期的作品中沒有像其早期那樣嚴格遵守小說結構的成規。由於最近發現的一部他的小說的結構像他早期的作品一樣嚴格地遵守了那些成規，因此該作品一定創作於他的早期。

   上面論述所依據的假設是：

   A. 作家在其創作晚期比早期更不願意打破某種成規

   B. 隨著創作的發展，作家日益意識不到其小說結構的成規

   C. 在其職業生涯晚期，該作家是其時代惟一有意打破小說結構成規的作家

   D. 作家在其創作生涯的晚期沒有寫過任何模仿其早期作品風格的小說

2. 儘管新製造的國產汽車的平均油效仍低於新製造的進口汽車，但它在1996年到1999年間卻顯著地提高了。自那以後，新製造的國產汽車的平均油效沒再提高，但新製造的國產汽車與進口汽車在平均油效上的差距卻逐漸縮小。

如以上論述正確，那麼基於此也一定正確的一項是：

A. 新製造的進口汽車的平均油效從1999年後逐漸降低

B. 新製造的國產汽車的平均油效從1999年後逐漸降低

C. 1999年後製造的國產汽車的平均油效高於1999年後製造的進口汽車的平均油效

D. 1996年製造的進口汽車的平均油效高於1999年製造的進口汽車的平均油效

3. 在1970年到1980年之間，世界工業的能源消耗量在達到頂峰後下降，1980年雖然工業總產出量有顯著提高，但工業的能源總耗用量卻低於1970年的水平。這說明，工業部門一定採取了高效節能措施。

   最能削弱上述結論的是：

   A. 1970年前，許多工業能源的使用者很少注意節約能源

   B. 20世紀70年代一大批能源密集型工業部門的產量急劇下降

   C. 工業總量的增長1970年到1980年間低於1960至1970年間的增長

   D. 20世紀70年代，許多行業從使用高價石油轉向使用低價的替代物

4. 在某國，10年前放鬆了對銷售拆鎖設備的法律限制後，盜竊案發生率急劇上升。因為合法購置的拆鎖設備被用於大多數盜竊案，所以重新引入對銷售該設備的嚴格限制將有助於減少該國的盜竊發生率。

最有力地支持以上論述的一項是：

A. 該國的總體犯罪率在過去10年中急劇增加了

B. 5年前引進的對被控盜竊的人更嚴厲的懲罰對該國盜竊率沒什麼影響

C. 重新引入對拆鎖設備的嚴格限制不會阻礙執法部門對這種設備的使用

D. 在該國使用的大多數拆鎖設備是易壞的，通常會在購買幾年後損壞且無法修好

5. 如果一個人增加日進餐次數並且不顯著增加所攝入的食物總量，那麼他的膽固醇水平會有顯著下降。然而，大多數增加日進餐次數的人同時也攝入了更多的食物。

   上面陳述支持的觀點是：

   A. 對大多數人而言，膽固醇的水平不受每天吃的食物量的影響

   B. 對大多數人而言，每頓飯吃的食物的量取決於吃飯的時間

   C. 對大多數人而言，增加每天吃飯的次數將不會導致膽固醇水平顯著下降

   D. 對大多數人而言，每天吃飯的總量不受每天吃飯的次數影響

6. 政府對基本商品徵收的一種稅是對出售的每一罐食用油徵稅兩分錢。稅務紀錄顯示，儘管人口數量保持穩定且稅法執行有力，食用油的稅收額在稅法生效的頭兩年中還是顯著下降了。

   如果正確，最有助於解釋食用油的稅收額下降的一項是：

   A. 很少家庭在加稅後開始生產他們自己的食用油

   B. 商人在稅法實施後開始用比以前更大的罐子售油

   C. 在食用油稅實行後的兩年，政府開始在許多其他基本商品上征稅

   D. 食用油罐傳統上被用作結婚禮物，稅法實施後，用食用油做禮物增多了

7. 與新疆的其他城市一樣，庫爾勒直至20世紀80年代初物價都是很低的，自它成為新疆的石油開採中心以後，它的物價大幅上升，這種物價上漲可能來自這場石油經濟，這是因為新疆那些沒有石油經濟的城市仍然保持著很低的物價水平。

最準確地描述了上段論述中所採用的推理方法的一項是：

A. 鑒於條件不存在的時候現像沒有發生，所以認為條件是現像的一個原因

B. 鑒於有時條件不存在的條件下現像也會發生，所以認為條件不是現像的前提

C. 由於某一特定事件在現像發生前沒有出現，所以認為這一事件不可能引發現像

D. 試圖說明某種現像是不可能發生的，而某種解釋正確就必須要求這種現像發生

8. 針對某種潰瘍，傳統療法可在6個月內將44%的患者的潰瘍完全治愈。針對這種潰瘍的一種新療法在6個月的試驗中使治療的80%的潰瘍取得了明顯改善，61%的潰瘍得到了痊愈。由於該試驗只治療了那些病情比較嚴重的潰瘍，因此這種新療法顯然在療效方面比傳統療法更顯著。

為更好地對比兩種療法的效果，還需要補充的證據是：

A. 這兩種療法使用的方法有何不同

B. 這兩種療法的使用成本是否存在很大差別

C. 在6個月中以傳統療法治療的該種潰瘍的患者中，有多大比例取得了明顯改善

D. 在參加6個月的新療法試驗的患者中，有多大比例的人對康復的比例不滿意

II. Verbal Reasoning (English) (6 questions)

In this test, each passage is followed by three statements (the questions). You have to assume what is stated in the passage is true and decide whether the statements are either:

(A) True: The statement is already made or implied in the passage, or follows logically from the passage.

(B) False: The statement contradicts what is said, implied by, or follows logically from the passage.

(C) Can't tell: There is insufficient information in the passage to establish whether the statement is true or false.

Passage 1 (Question 9 to 11)

The clinical guidelines in asthma therapy have now moved towards anti-inflammatory therapy - and away from regular bronchodilator therapy - for all but the mildest asthmatics. This is now being reflected in prescribing patterns. In the U.S., combined prescription volumes of the major bronchodilators peaked in 1991 (having risen slowly in the preceding years), though they still account for around half of the 65 million asthma prescriptions there. During the same period, prescriptions for inhaled steroids have doubled, but still account for less than 10% of asthma prescriptions in the U.S.

9. Only mild cases of asthma can be helped by anti-inflammatory therapy.

10. Use of bronchodilators has been increasing since 1991.

11. Doctors are reluctant to treat asthma with inhaled steroids for fear of potential side-effects.

Passage 2 (Question 12 to 14)

Relations between Sweden and the European Community had always been restricted in scope by Sweden's traditional neutrality and for many years any suggestion of Community membership was out of the question. But the upheavals in Eastern Europe in the early 1990s gradually led to the conclusion that membership of the EC was no longer incompatible with its neutral stance. People came to the conclusion that Sweden has already taken over a large part of the Community rules and began to weigh up the pros and cons of membership along the lines sought by Austria.

12. Political changes in Eastern Europe led to a change in relations between Sweden and the European Community.

13. The European Community rejected Sweden's application for membership because of its neutrality.

14. After abandoning its policy of neutrality, Sweden applied to join the European Community.

III. Data Sufficiency Test (8 questions)

In this test, you are required to choose a combination of clues to solve a problem.

15. Set 1: 13, 10, 8, 2

    Set 2: 13, 11, 3, 2, x

    Is the median of the numbers in set 1 above, equal to the median of the numbers in set 2?

    (1) $8 < x < 10$

    (2) The sum of numbers in set 2 is 38

    A. statement 1 alone is sufficient, but statement 2 alone is not sufficient to answer the question
    B. statement 2 alone is sufficient, but statement 1 alone is not sufficient to answer the question
    C. both statements taken together are sufficient to answer the question, but neither statement alone is sufficient
    D. each statement alone is sufficient
    E. statements 1 and 2 together are not sufficient, and additional data is needed to answer the question

16. What is the average of x, y, and z?

(1) $2x + y + 4z = 23$

(2) $3x + 4y + z = 22$

A. statement 1 alone is sufficient, but statement 2 alone is not sufficient to answer the question

B. statement 2 alone is sufficient, but statement 1 alone is not sufficient to answer the question

C. both statements taken together are sufficient to answer the question, but neither statement alone is sufficient

D. each statement alone is sufficient

E. statements 1 and 2 together are not sufficient, and additional data is needed to answer the question

17. Are p and q both greater than zero?

(1) $p - q > 0$

(2) $p > q$

A. statement 1 alone is sufficient, but statement 2 alone is not sufficient to answer the question

B. statement 2 alone is sufficient, but statement 1 alone is not sufficient to answer the question

C. both statements taken together are sufficient to answer the question, but neither statement alone is sufficient

D. each statement alone is sufficient

E. statements 1 and 2 together are not sufficient, and additional data is needed to answer the question

18. What is the value of $3x^2 + 2x - 1$?

(1) $x^2 + 2x = 0$

(2) $x = -2$

A. statement 1 alone is sufficient, but statement 2 alone is not sufficient to answer the question

B. statement 2 alone is sufficient, but statement 1 alone is not sufficient to answer the question

C. both statements taken together are sufficient to answer the question, but neither statement alone is sufficient

D. each statement alone is sufficient

E. statements 1 and 2 together are not sufficient, and additional data is needed to answer the question

19. What is the ratio of *AB* to *BC* in the rectangular figure below?

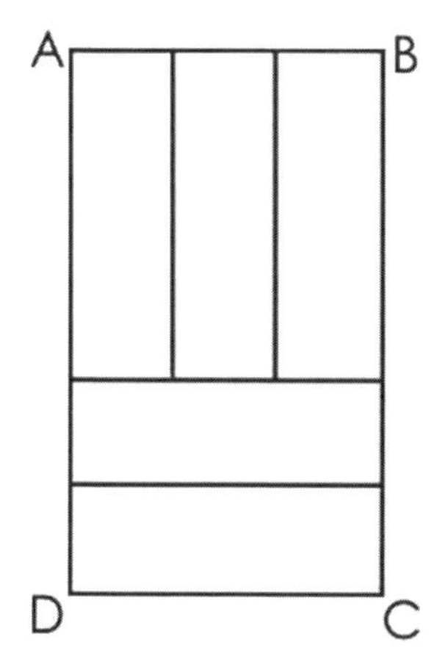

(1) The perimeter of *ABCD* is 32.

(2) All the small rectangles have the same dimensions.

A. statement 1 alone is sufficient, but statement 2 alone is not sufficient to answer the question

B. statement 2 alone is sufficient, but statement 1 alone is not sufficient to answer the question

C. both statements taken together are sufficient to answer the question, but neither statement alone is sufficient

D. each statement alone is sufficient

E. statements 1 and 2 together are not sufficient, and additional data is needed to answer the question

20. A team of workers including Tom and Dick work in the same office according to a schedule that ensures that exactly two team members will be present at a given time, and that in the course of the week all the team members work an equal number of hours. What is the probability that a visitor to the office who doesn't know the schedule arrives to find both Tom and Dick in the office?

(1) The team has three members.

(2) Tom and Dick worked together for the whole of the previous day.

A. statement 1 alone is sufficient, but statement 2 alone is not sufficient to answer the question

B. statement 2 alone is sufficient, but statement 1 alone is not sufficient to answer the question

C. both statements taken together are sufficient to answer the question, but neither statement alone is sufficient

D. each statement alone is sufficient

E. statements 1 and 2 together are not sufficient, and additional data is needed to answer the question

21. In triangle ABC, $AB = x$, $BC = y$, and $CA = y - 4$. Which of the three angles of triangle ABC is the smallest?

(1) $y = x + 5$

(2) $x = 6$

A. statement 1 alone is sufficient, but statement 2 alone is not sufficient to answer the question

B. statement 2 alone is sufficient, but statement 1 alone is not sufficient to answer the question

C. both statements taken together are sufficient to answer the question, but neither statement alone is sufficient

D. each statement alone is sufficient

E. statements 1 and 2 together are not sufficient, and additional data is needed to answer the question

22. A man shared his lottery winnings of $15,000 with his wife and two children. How much did his wife receive?

(1) He gave his children $2,000 each.

(2) The wife received $1,000 less than her husband and three thousand more than each of the children.

A. statement 1 alone is sufficient, but statement 2 alone is not sufficient to answer the question

B. statement 2 alone is sufficient, but statement 1 alone is not sufficient to answer the question

C. both statements taken together are sufficient to answer the question, but neither statement alone is sufficient

D. each statement alone is sufficient

E. statements 1 and 2 together are not sufficient, and additional data is needed to answer the question

IV. Numerical Reasoning (5 questions)

Each question is a sequence of numbers with one or two numbers missing. You have to figure out the logical order of the sequence to find out the missing number(s).

23. 93, 114, 136, 159, ( )

A.180　　B.183
C.185　　D.187

24. 20, 31, 43, 56, ( )

A.68　　B.72
C.80　　D.70

25. 918, 818, 717, 615, ( )

A.495　　B.508
C.512　　D.451

26. 7, 9, 62, 557, ( )

A.1 537　　B.30 513
C.34 533　　D.33 503

27. 2, 5, 8, 11, 14, ( )

A.17　　B.16
C.18　　D.19

V. Interpretation of Tables and Graphs (8 questions)

This is a test on reading and interpretation of data presented in tables and graphs.

Graph 1 (Question 28 to 30)

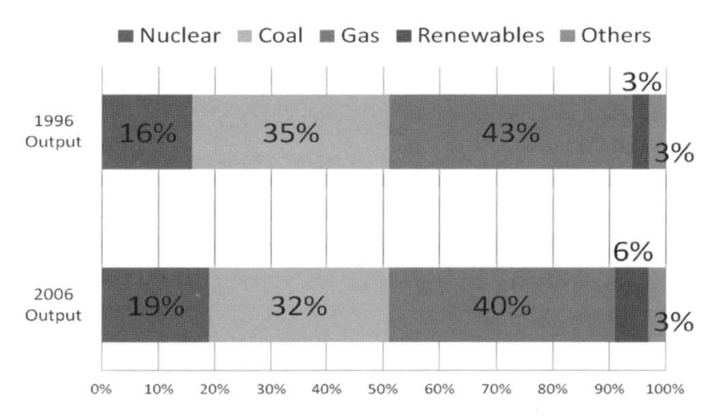

28. If output for Nuclear in 2006 was twice that for Coal in 1996 when total output was 200TWh, what was the output for Nuclear in 2006?

    A. 140TWh B. 400TWh C. 64TWh D. 96TWh

29. Now if instead, Renewables output doubled to 18TWh between 1996 and 2006, approximately what was the Gas output in 1996?

    A. 129TWh B. 86TWh C. 120TWh D.110TWh

30. If total output was 200TWh in 1996 and 320TWh in 2006, what was Gas output in 2001?

    A.107TWh B. 133TWh C. 120TWh D. Cannot say

Graph 2 (Question 31 to 33)

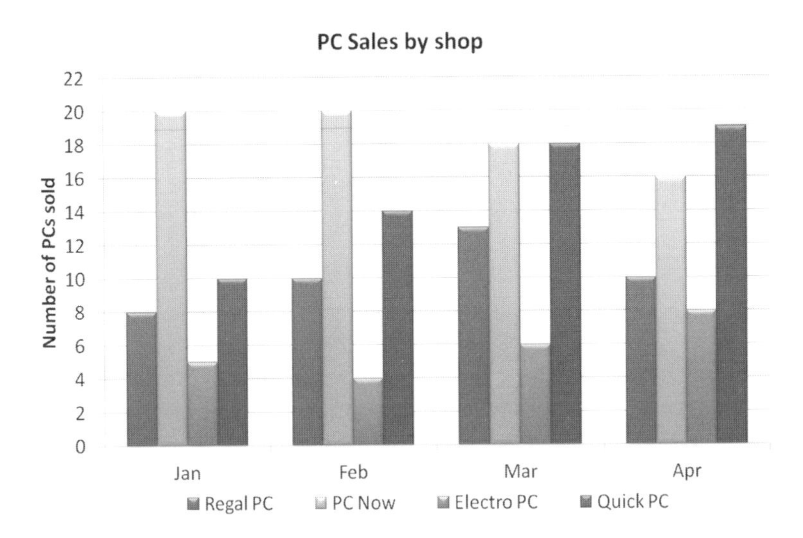

31. Which month showed the largest total decrease in PC sales over the previous month?

A. Jan B. Feb C. Mar D. Apr

32. Approximately what percentage of Regal PC's sales were made in April?

A. 24% B. 21% C. 22% D. 28%

33. If the average profit made on each PC sold by Quick PC over all four months was £62 what was the total profit over the four months for this shop?

A. £3,782 B. £3,144 C. £3,857 D. Cannot say

Graph 3 (Question 34 to 35)

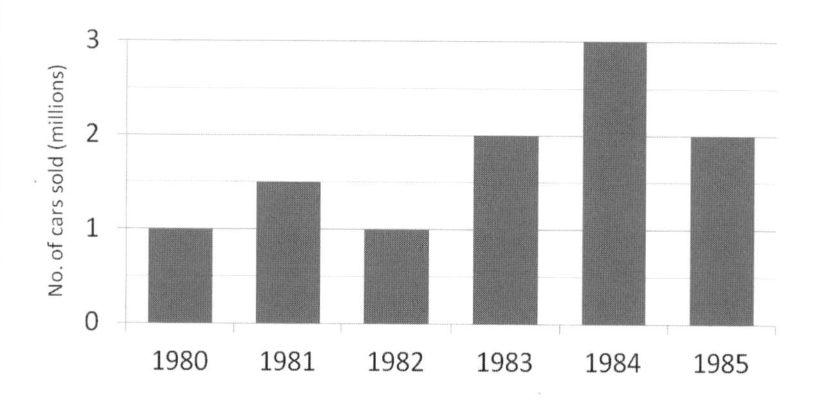

34. According to the information in the graph, what year saw a decrease in the number of cars sold?

A. 1980 B. 1981 C. 1982 D. 1983 E. 1984

35. According to the information in the graph, in what two years were the number of cars sold closest to equal?

A. 1980 and 1981
B. 1981 and 1982
C. 1981 and 1983
D. 1983 and 1985
E. 1984 and 1985

ANSWERS

I. 演繹推理 (8題)

1. 答案：D。排除法。要選擇論述所依據的假設，那麼這個假設應該與題幹中論述的說法一致。A項的說法與題幹第一句話就矛盾，題幹中只說明作家的晚期小說沒有嚴格的遵守小說結構的成規，並不意味着作家就意識不到這種成規的存在，B也不能選；同時題幹中並沒有說明同時代其他作家的情況，故C也是不能選的。

2. 答案：A。1999年之後，國產汽車的平均油效沒有變化，但是國產汽車與進口汽車的平均油效之間的差距在不斷減小，這只能說明進口洌車的平均油效在不斷降低，B錯誤而A正確；從題幹的第一句話可以判定，到1999年，國產洌車的平均油效是低於進口汽車的，故C不正確；D項在題幹中沒有提到。

3. 答案：D。本題的主要結論是工業部採取高效的節能措施導致了工業能源總耗用量的降低。要削弱這個結論，就要找出另外一個導致能源總耗用量降低的量由。D項用低價替代物替代高價石油，促使石油的消耗量的降低，剛好合符條件，所以當選。

4. 答案：B。要降低盜竊發生率，需要兩個途徑一個是加大對盜竊犯的懲罰力度，二是重新引入對銷售該設備的限制。題幹要求強化第二種辦法的作用，那麼就要弱化第一種方法的作用，也就是B項的內容。

5. 答案：C。膽固醇水平下降需要同時具備兩個條件：第一是增加每天進餐的次數，第二個是每天攝入的食物總量不變，也就是說要“少吃多餐”。低是由於大多數人在多餐的同時也“多吃”了，所以他們的膽固醇水平並不能降低。所以，A、B、D項說法有誤，C項正確。

6. 答案：B。本題實際上是要求找出導致食用油稅收額下降的原因。顯然，C，項可以排除；用做禮物的食用油也是從市場上流通而得來的，因此也要繳納稅收，故D項不正確；A項條件下，食用油的稅收只會增多而不會減少；由於使用了更大的罐子，在食用油總量不變的情況下，所能出產的總罐數就會減少，由於征稅是按照罐數徵收的，所以在這種條件下，稅收就會減少。

7. 答案：A。當成為開採中心這一條件不存在的時候，庫爾勒的物價上漲的現象並沒有發生，故B項不正確；根據C項的意思，“某一特定事件”指成為開採中心這件事，“現象”就是指物價上漲，成為開採中心之後物價沒有上漲，所以成為開採中心不能引起物價上漲，這種說法明顯不正確；而D項的說法過於模糊，指代不清，故也不選。

8. 答案： C。所需要的證據是為了說明新療法在“療效”方面勝過傳統療法，所以證據必須是和“療效”有關的，A、B可以排除。D項所表達的“人們對新療法在改善和使患者痊癒兩方面的作用，而未提到傳統療法的治癒率，所以在文中可以增加，也就是C的內容。

II. Verbal Reasoning (English) (6 questions)

9. B (False): Only mild cases of asthma can be helped by anti-inflammatory therapy.

10. B (False): Use of bronchodilators has been increasing since 1991.

11.C (Can’t Say): Doctors are reluctant to treat asthma with inhaled steroids for fear of potential side-effects.

12.A (True): Political changes in Eastern Europe led to a change in relations between Sweden and the European Community.

13.B (False): The European Community rejected Sweden’s application for membership because of its neutrality.

14.B (False): After abandoning its policy of neutrality, Sweden applied to join the European Community.

III. Data Sufficiency Test (8 questions)

15. B　16. C　17. E
18. B

19. B　20. A　21. A
22. B

IV. Numerical Reasoning (5 questions)

23. B　24. D　25. C
26. C　27. A

V. Interpretation of Tables and Graphs (8 questions)

28. A: Coal in 1996 was 35% of 200TWh: 70TWh. So Nuclear in 2006 is twice this: 140TWh.

29. A: So in 2006, Renewables was 18TWh, which was double what it was in 1996: 9TWh. Therefore, if 9TWh is 3% of the total (according to chart), total in 1996 was 300TWh. Gas we are told made up 43% in 1996, i.e. 0.43 x 300TWh = 129TWh.

30. D: No data is given, or can be implied, for 2001 in the question.

31. D: We have to sum for every month, even though we have no data for Dec and therefore Jan cannot be the answer.

    Jan: 8 + 20 + 5 + 10 = 43

    Feb: 10 + 20 + 4 + 14 = 48

    Mar: 13 + 18 + 6 + 18 = 55

    Apr: 10 + 16 + 8 + 19 = 53

    April was the only month where sales dropped.

32. A: Total PC sales for Regal = 8 + 10 + 13 + 10 = 41. April = 10. So 10 ÷ 41 = 24.4%.

33. A: Total sales by Quick PC over the four months = 10 + 14 + 18 + 19 = 61. So 61 x £62 = £3,782

34. C

35. D

能力傾向測試
模擬測驗（二）

限時四十五分鐘

I. 演繹推理(8題)

1. 亞清是今年才進入仁一大學讀書，他以往的成績不錯。亞清最喜歡讀數學，但是他在仁一大學所讀的就不是他喜歡的科目；亞清認為仁一大學應該讓他修讀數學，但修讀數學原來要有相當的成績才可修讀。亞清讀書首先是講求興趣，沒有興趣的科目，他會讀得很辛苦。亞清認為仁一大學應該有一些科目，可以讓成績不足的學生去修讀。由此可推論：

   A. 亞清在大學讀書成績不好。

   B. 亞清讀數學，不會感到辛苦。

   C. 亞清在仁一大學沒有修讀數學。

   D. 大學有一些科目可以讓成績不足的學生修讀，只是亞清不知而已。

2. 李先生住在木本城的西區，木本城最大的商場在東區。李先生是木本城最大商場的僱員，但他只需要每星期上班三天。李先生只在上班日子才去東區，因為木本城東區與西區之間的車程有一個小時，所以李先生覺得舟車時間十分浪費；如果可以選擇，他會選擇在原區工作，而他剛辭了職。由此可推論：

   A. 李先生會在西區工作。

   B. 李先生每星期只有三天在木本城的東區出現。

   C. 李先生覺得木本城的東區與西區距離很遠。

   D. 李先生有四天是在木本城西區活動。

3. 王村有一條村規：凡王村的人不可以嫁給王村的人。觸犯這一村規的村民都要離開王村。王小華是王村的村民，他與李小明結婚已經一年多；李小明是王小華青梅竹馬的好朋友；兩大長大後，情投意合就結了婚。結婚後，他倆住在王村村口的健華大廈，早出晚歸，生活也算不錯。由此可推論：

   A. 李小明不是王村的村民。

   B. 健華大廈不在王村村內。

   C. 李小明和王小華自小在王村認識。

   D. 王小華沒有遵守王村村規。

4. 熊大公司做的是中、港、台貿易的生意；公司的總部在香港，不過僱員最多來自中國大陸。現時，熊大公司約有僱員5000人，而在下個月將在中國內地招聘500員工。熊大公司的管理階層大多數來自中國內地，其他的都是香港人。由此可推論：

   A. 熊大公司的台灣市場最小。

   B. 下個月後，熊大公司將多了500名員工。

   C. 熊大公司中來自中國內地的員工都在香港入住熊大公司宿舍。

   D. 有香港人在熊大公司的管理層。

5. 每次開會，亞陳一定坐在我的右邊，凌秘書一定坐在我的左邊，紅紅有時會坐在我的後邊，有時坐在亞陳的右邊。小敏每次開會都遲到，公司規定，開會遲到的，如果會議室內沒有安排座位的就坐在最近僧議室門口的座位。柱明每次開會負責控制秩序，雖然大會已為他安排座位，但他通常不在座位上。今天上午開會，坐在最近會議室門口座位的不是小敏。由此可推論：

A.今天的會議只有六個人參加。

B. 小敏沒有遲到。

C. 在今天的會議中，亞陳和凌秘書都坐在我的旁邊。

D. 在今天的會議中，紅紅坐在我的後面。

6. 何先生真是有福氣：他是何家村唯一一個四代同堂的人；他有四個兒子，一個女兒。除了最小的兒子外，其他的已經結了婚。大兒子在外國工作，將在今年夏天回來度假。何先生有六個內孫，兩個外孫，全是男孫。六個內孫中有兩個已成家立室，其中一個內孫就在上個月做了父親。何先生現時最大的心願，就是他最小的兒子可以成家立室。由此可推論：

A. 何家村中除了何先生一家之外，其他的家庭人口都很少。

B. 何先生最小的兒子即將成家立室。

C. 何先生的大兒子移民外國。

D. 何先生的曾孫也是姓何。

7. 歐太太約了她的姊姊到城南最大的超級市場辦年貨。歐太太比她的姊姊傳統，對中國賀年食物有多一點的要求。歐太太打算今次辦年貨，至少花上五千元，希望家人感到過年有個豐盛的感覺。歐太太的姊姊覺得「過年如過日」，不必只為過年辦年貨，平日有需要也可以購物，所以沒有定下購物的金額。她倆都喜歡到城南這間超級市場買東西。這間超級市場的老闆交遊廣闊，有時會給予相熟顧客多一點折扣。由此可推論：

A. 歐太太姊姊打算比歐太太少買一點東西。

B. 歐太太要花上五千元辦年貨。

C. 到城南的超級市場去買東西，歐太太比她的姊姊多一點折扣。

D. 歐太太打算多買一點傳統的東西回來過年。

8. 洪先生對於本城的公園都很稔熟。本城只有三個公園；最大的一個在城西，而最小一個在城東。而洪先生最喜歡流連的是不大不小的公園。因為這個公園最接近他所居住的地方。他每天早上及黃昏都會經過這個公園。他覺得最大的公園設施最好，但最人工化，而且座落在人口最密集的區域。由此可推論：

A. 不大不小的公園座落在城中。

B. 洪先生每天都要經過不大不小的公園才可回到家。

C. 洪先生最不認識城東的公園。

D. 本城城西人口最密集。

II. Verbal Reasoning (English) (6 questions)

In this test, each passage is followed by three statements (the questions). You have to assume what is stated in the passage is true and decide whether the statements are either:

(A) True: The statement is already made or implied in the passage, or follows logically from the passage.

(B) False: The statement contradicts what is said, implied by, or follows logically from the passage.

(C) Can't tell: There is insufficient information in the passage to establish whether the statement is true or false.

Text 1 (Question 9 to 11)

Entrepreneurs running small firms play a vital role in ensuring a healthy economy, not just from a business perspective, but also in social, educational and political terms. They compete with the large businesses that would otherwise dominate the markets and are key providers of new jobs. Smaller businesses are able to accommodate working patterns tailored to the employee's needs. They are, therefore, valuable sources of employment for the large number of people with family responsibilities who wish to remain part of the labour market but are unable, because of domestic commitments, to take up full-time employment.

9. Entrepreneurs tend not to compete with large organizations.

   A. True

   B. False

   C. Can't tell

10. Large businesses do not want to accommodate an employee's individual employment needs.

    A. True

    B. False

    C. Can't tell

11. Small firms run by entrepreneurs provide no benefits for the community.

    A. True

    B. False

    C. Can't tell

Text 2 (Question 12 to 14)

Advertising and selling books via Internet sites is becoming more popular with traders. It costs less to publicize a book on the Internet than by traditional methods, and as books are stored in warehouses prior to being dispatched to customers, overheads are lower than those of shops. True, the price war on the Internet is likely to put pressure on royalties, with publishers demanding that they be calculated not on the cover prices of books but on the prices actually received for them. However, these discounts will be greatest on best-sellers, rather than other books.

12. The consumer demand for books sold on the Internet is increasing.

    A. True

    B. False

    C. Can't tell

13. The cost of placing an advertisement for a book on the Internet is less than other methods of marketing.

    A. True

    B. False

    C. Can't tell

14. Internet bookstores offer their biggest discounts on less popular books.

    A. True

    B. False

    C. Can't tell

III. Data Sufficiency Test

15. In case Sue sits between Pete and Harry, then Harry sits between Sue and Mike. Harry won't be there unless Sue sits next to Mike. Hence, Sue will not sit between Pete and Harry.

    Apart from the above mentioned statements, what additional premises are assumed by the author of this argument?

    Mike sits next to Sue if no one sits between them.

    If Sue sits between Pete and Harry, then Sue sits between Harry and Pete.

    If Harry isn' t there, then he doesn' t sit next to Mike.

    A. I and II only

    B. I and III only

    C. II and III only

    D. I, II and III

    E. None of the above

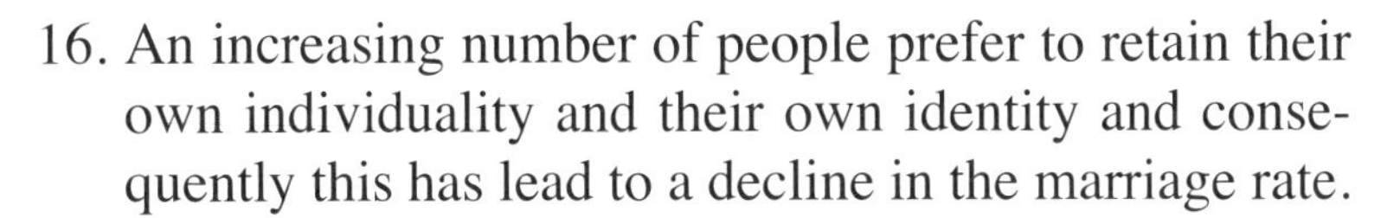
16. An increasing number of people prefer to retain their own individuality and their own identity and consequently this has lead to a decline in the marriage rate.

Which among the following assumptions are used in the above premises?

When a person is married, he or she loses his or her own identity and is no longer accountable to himself or herself.

Married persons do not find contentment as opposed to unmarried people.

There has been a steady increase in the divorce rate.

A. I only

B. II only

C. III only

D. I and II only

E. I, II, and III

Questions 17 to 19

Three girls Joan, Rita, and Kim and two boys Tim and Steve are the only dancers in a dance program, which consists of six numbers in this order: One a duet; two a duet; three a solo; four a duet; five a solo; and six a duet.

None of the dancers is in two consecutive numbers or in more than two numbers.

The first number in which Tim appears is the one that comes before the first number in which Kim appears.

The second number in which Tim appears is one that comes after the second number in which Kim appears.

17. Which among the following is a complete and accurate list of those numbers that could be the last one in which Kim performs?

    A. Three

    B. Four

    C. Five

    D. Three, Four

    E. Four, Five

18. Rita must perform only in duets if

A. Kim is in number two

B. Kim is in number five

C. Tim is in number one

D. Tim is in number two

E. Tim is in number six

19. In case Steve is in number five, number four must consist of

A. two women

B. two men

C. Tim and a woman

D. Rita and a man

E. Kim and a man

20. A survey recently conducted revealed that marriage is fattening. The survey found that on an average, women gained 23 pounds and men gained 18 pounds during 13 years of marriage. The answer to which among the following questions would be the most appropriate in evaluating the reasoning presented in the survey ?

A. Why is the time period of the survey 13 years, rather than 12 or 14 ?

B. Did any of the men surveyed gain less than 18 pounds during the period they were married ?

C. How much weight is gained or lost in 13 years by a single people of comparable age to those studied in the survey ?

D. When the survey was conducted were the women as active as the men?

E. Will the gains seen in the study be retained over the lifetimes of the surveyed persons?

21. It is popularly believed that teachers are more or less indifferent about the microcomputer technology. This assumption is false, or at least dated. A survey recently conducted indicated that 80 percent of the 7,000 surveyed teachers revealed a high level of interest in microcomputers.

Among the following statements which would most damage the above argument if proved to be true?

A. There was no attempt made in the survey to ascertain whether the surveyed teachers had any previous exposure to microcomputers.

B. Teachers interested in microcomputer technology were more likely to complete and return the questionnaires than others.

C. Irrespective of their subject area, their expertise and their teaching experience questionnaires were received by the teachers.

D. After the survey results were tabulated there have been many developments in the applications of microcomputer technology.

E. A company manufacturing and selling microcomputers conducted the survey.

22. Some lawyers are of the view that the observation of the intrinsic qualities of pornography in any composition depends on literary criticism and hence it is a matter of opinion. It is rather odd, though, that in a legal connection, serious critics themselves quite often behave as if they believed criticism to be a matter of opinion. Why be a critic - and teach in universities - in case criticism involves nothing but uttering capricious and arbitrary opinions ?

The above discussion would be weakened if it is pointed out that:

A. literary critics are of the opinion that nothing is pornographic.

B. lawyers believe that the observance of the qualities of pornography is a matter of opinion, as literary critics are not in agreement in this regard.

C. literary critics are not legal authorities.

D. literary critics should not concern themselves with deciding what is pornographic.

E. literary critics in the teaching profession at the university level are init only for the money.

IV. Numerical Reasoning (4 questions)

Each question is a sequence of numbers with one or two numbers missing. You have to figure out the logical order of the sequence to find out the missing number(s).

23. 86, ? , 79, 75, 72, 68

A.82 B.80 C.85 D.84

24. 11, 19, ?, 41, 55

A.31 B.29 C.26 D.39

25. 44, 31, 45, 30, 46, ( )

A47 B.29 C.48 D.28

26.

8	?	6	9	7
5	7	3	6	4

A.16 B.14 C.11 D.10

V. Interpretation of Tables and Graphs (8 questions)

Directions

This is a test on reading and interpretation of data presented in tables and graphs.

Population Statistics for 2007

Country	Population (millions)	5-14 year olds (% of population)	15-24 year olds (% of population)	Employed (% of population)
A	81.5	11.1	11.7	60
B	58.2	12.9	13.0	51
C	39.6	11.0	16.3	59
D	10.4	15.8	14.1	57

Percentage of the Employed Population using Different Modes of Transport to Travel to Work (2007 data)

Country	PRIVATE VEHICLE			PUBLIC VEHICLE			Other
	Car/Van	Motor Cycle	Pedal Cycle	Bus / Coach	Rail	Boat / Plane	
A	42	20	9	19	6	2	2
B	12	11	24	9	31	2	11
C	33	8	6	14	2	4	6
D	53	7	16	2	13	3	6

27. Which country has the same percentage of people using private and public vehicles to get to work?

A. Country A B. Country B
C. Country C D. Country D

28. In which country is there the biggest difference between the numbers of people in the 5 – 14 and 15 – 24 year old categories?

A. Country A B. Country B
C. Country C D. Country D

29. Approximately how many people in country B travel to work by motorcycle?

A. 3 million B. 5 million
C. 7 million D. 9 million

30. What is the approximate difference between the number of people taking public and private vehicles to work in country B ?

A. 1 million B. 1.5 million
C. 3 million D. Cannot say

Population Statistics for 2007

Country	Population (millions)	5-14 year olds (% of population)	15-24 year olds (% of population)	Employed (% of population)
A	81.5	11.1	11.7	60
B	58.2	12.9	13.0	51
C	39.6	11.0	16.3	59
D	10.4	15.8	14.1	57

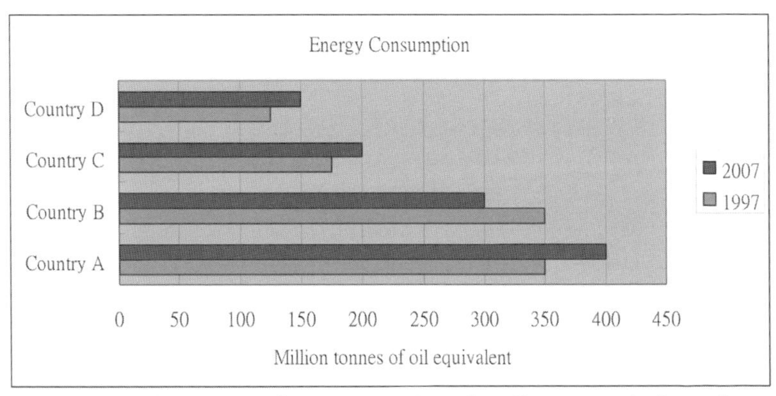

Sources of Energy Consumption for Country A (in %)

31. In country A, which source of energy had the largest proportional change between 1987 and 1997?

A. Coal B. Petroleum C. Natural Gas D.Other

32. For country A in 1987 how much of the total energy consumption was provided for by coal?

A. 85 Million tonnes of oil equivalent

B. 95 Million tonnes of oil equivalent

C. 105 Million tonnes of oil equivalent

D. 115 Million tonnes of oil equivalent

33. Which country showed the greatest percentage change in total energy consumption between 1987 and 1997?

A. Country A B. Country B
C. Country C D. Country D

34. Country C anticipates that energy consumption per million population will increase at a rate of 10% a year from 1997. If the energy consumption of country B remains constant, how many years will it be before country C's consumption exceeds that of country B ?

A. 3 years B. 4 years
C. 5 years D. 6 years

ANSWERS

演繹推理 (8題)

1.C 2. B 3. A 4. D 5. C 6. D 7. B 8. D

Verbal Reasoning

9.B 10. C 11. B 12 C 13. A 14. B

Data Sufficiency

15. B 16. A 17. E 18. D
19. A 20. C 21. B 22. B

Numerical Reasoning

23. A 24. B 25. B 26. D

Interpretation of Tables and Graphs

27. C 28. C 29. A 30. B
31. B 32. C 33. D 34. C

能力傾向測試
模擬測驗（三）

限時四十五分鐘

I. 演繹推理(8題)

請根據以下短文的內容，選出一個或一組推論。請假定短文的內容都是正確的。

1. 當一名司機被懷疑飲用了過多的酒精時，檢驗該司機走直線的能力與檢驗該司機血液中的酒精水平相比，是檢驗該司機是否適於駕車的更可靠的指標。

   如果正確，能最好地支持上述觀點的一項是：

   A. 觀察者們對一個人是否成功地走了直線不能全部達成一致

   B. 用於檢驗血液酒精含量水平的測試是準確、低成本和易於實施的

   C. 一些人在血液酒精含量水平很高時，還可以走直線，但卻不能完全駕車

   D. 由於基因的不同和對酒精的抵抗能力的差別，一些人血液酒精含量水平很高時仍能正常駕車

2. 桌上放著紅桃、黑桃和梅花三種牌，共20張，

［1］桌上至少有一種花色的牌少於6張

［2］桌上至少有一種花色的牌多於6張

［3］桌上任意兩種牌的總數將不超過19張

上述論述中正確的是：

A. ［1］、［2］

B. ［1］、［3］

C. ［2］、［3］

D. ［1］、［2］和［3］

3. 小王：從舉辦奧運會的巨額耗費來看，觀看各場奧運比賽的票價應該要高得多。是奧運會主辦者的廣告收入降低了票價。因此，奧運會的現場觀眾從奧運會拉的廣告中獲得了經濟利益。

小李：你的說法不能成立。誰來支付那些看來導致奧運會票價降低的廣告費用？到頭來還不是消費者，包括作為奧運會現場觀眾的消費者。因為廠家通過提高商品的價格把廣告費用攤到了消費者的身上。

下列能夠有力地削弱小李對小王反駁的一項是：

A. 奧運會的票價一般要遠高於普通體育比賽的票價

B. 奧運會的舉辦帶有越來越濃的商業色彩，引起了普遍的不滿

C. 利用世界性體育比賽做廣告的廠家越來越多，廣告費用也越來越高

D. 各廠家的廣告支出總體上是一個常量，只是在廣告形式上有所選擇

4. 北京市是個水資源嚴重缺乏的城市，但長期以來水價格一直偏低。最近北京市政府根據價值規律擬調高水價，這一舉措將對節約使用該市的水資源產生重大的推動作用。

若上述結論成立，下列哪些項必須是真的：

［1］有相當數量的用水浪費是因為水價格偏低造成的

［2］水價格的上調幅度足以對浪費用水的用戶產生經濟壓力

［3］水價格的上調不會引起用戶的不滿

A.［1］、［2］

B.［1］、［3］

C.［2］、［3］

D.［1］、［2］和［3］

5. 西方發達國家的大學教授幾乎都是得到過博士學位的。目前，我國有些高等學校也堅持在招收新教師時，有博士學位是必要條件，除非是本校的少數優秀碩士畢業生留校。

從這段文字中可以推出的是：

A. 在我國，有些高等學校的新教師都有了博士學位

B. 在我國，有些高等學校得到博士學位的教師的比例在增加

C. 大學教授中得到博士學位的比沒有得到博士學位的更受學生歡迎

D. 在我國，大多數大學教授已經獲得了博士學位，少數正在讀在職博士

6. 環境學家關注保護瀕臨滅絕的動物的高昂費用，提出應通過評估各種瀕臨滅絕的動物對人類的價值，以決定保護哪些動物。此法實際不可行，因為，預言一種動物未來的價值是不可能的。評價對人類現在做出間接但很重要貢獻的動物的價值也是不可能的。

從這段文字中可以推出的是：

A. 保護對人類有直接價值的動物遠比保護有間接價值的動物重要

B. 保護沒有價值的瀕臨滅絕的動物比保護有潛在價值的動物更重要

C. 儘管保護所有瀕臨滅絕的動物是必須的，但在經濟上卻是不可行的

D. 由於判斷動物對人類價值高低的方法並不完善，在此基礎上做出的決定也不可靠

7. 研究人員對75個胎兒進行了跟蹤調查，他們中的60個偏好吸吮右手，15個偏好吸吮左手。在這些胎兒出生後成長到10到12歲時，研究人員發現，60個在胎兒階段吸吮右手的孩子習慣用右手；而在15個吸吮左手的胎兒中，有10個仍舊習慣用左手，另外5個則變成「右撇子」。

從這段文字中，不能推出的是：

A. 大部分人是「右撇子」

B. 人的偏側性隨著年齡的增長不斷改變

C. 大多數人的偏側性在胎兒時期就形成了

D. 「左撇子」可能變成「右撇子」，而「右撇子」難變成「左撇子」

8. 甲：兒時進行大量閱讀會導致近視眼——難以看清遠處景物。

乙：我不同意，近視眼與閱讀之間的關聯都來自以下事實：觀看遠處景物有困難的孩子最有可能選擇那些需要從近處觀看物體的活動，如閱讀。

乙對甲的反駁是通過：

A. 運用類比來說明甲推理中的錯誤

B. 指出甲的聲明是自相矛盾的

C. 說明如果接受甲的聲明，會導致荒謬的結論

D. 論證甲的聲明中某一現像的原因實際上是該現像的結果

II. Verbal Reasoning (English) (6 questions)

In this test, each passage is followed by three statements (the questions). You have to assume what is stated in the passage is true and decide whether the statements are either:

(A) **True:** The statement is already made or implied in the passage, or follows logically from the passage.

(B) **False:** The statement contradicts what is said, implied by, or follows logically from the passage.

(C) **Can't tell:** There is insufficient information in the passage to establish whether the statement is true or false.

Passage 1 (Question 9 to 11)

Buddhism was introduced to Japan from India via China and Korea around the middle of the sixth century. After gaining imperial patronage, Buddhism was propagated by the authorities throughout the country. In the early ninth century, Buddhism in Japan entered a new era in which it catered mainly to the court nobility. In the Kamakura period (1192-1338), an age of great political unrest and social confusion, there emerged many new sects of Buddhism offering hope of salvation to warriors and peasants alike. Buddhism not only flourished as a religion but also did much to enrich the country's arts and learning.

9. Buddhism was adopted by the court nobility at the urging of the emperor.

10. The introduction of Buddhism to Japan led to great political unrest and social confusion.

11. Buddhism replaced the Shinto religion which had previously been followed in Japan.

Passage 2 (Question 12 to 14)

In Japan, companies generally expect their employees to put in long hours of overtime. But it is difficult for women, who also have household chores to do and children to take care of, to work at the same pace as men, who are not burdened with such responsibilities. Many women inevitably opt for part-time jobs, which enable them to combine work and domestic duties. At present, 23% of all female salaried workers are part-timers and the ratio has been on the rise in recent years. Part-time work places women at a disadvantage. The wages of part-time workers are considerably lower than those of full-time employees, and part-time work tends to involve menial labour. Moreover, because salary and promotion in Japanese companies are often based on seniority, it is extremely difficult for women either re-entering the labour force or switching from part-time to full-time work to climb the ladder.

12. Japanese men do not share household chores and childcare with their wives.

13. A quarter of all part-time workers in Japan are female.

14. Part-time workers hold a low status in Japanese companies.

III. Data Sufficiency Test (8 questions)

In this test, you are required to choose a combination of clues to solve a problem.

15. Does $x^3 - x$ give a whole number when divided by 3?

    (1) x is a positive integer greater than 1.

    (2) $|x| > 0$

    A. statement 1 alone is sufficient, but statement 2 alone is not sufficient to answer the question

    B. statement 2 alone is sufficient, but statement 1 alone is not sufficient to answer the question

    C. both statements taken together are sufficient to answer the question, but neither statement alone is sufficient

    D. each statement alone is sufficient

    E. statements 1 and 2 together are not sufficient, and additional data is needed to answer the question

16. $x = 1.24d5$

If d is the thousandth's digit in the decimal above, what is the value of x when rounded to the nearest hundredth?

(1) $x < 5/4$

(2) $d < 5$

A. statement 1 alone is sufficient, but statement 2 alone is not sufficient to answer the question

B. statement 2 alone is sufficient, but statement 1 alone is not sufficient to answer the question

C. both statements taken together are sufficient to answer the question, but neither statement alone is sufficient

D. each statement alone is sufficient

E. statements 1 and 2 together are not sufficient, and additional data is needed to answer the question

17. Three teams, A, B and C each have five members. The three teams competed in a three-event competition. Five points were awarded to the winner of each event, four points to the second and three points to the third. (No points were awarded for lower positions). Which team won the competition?

(1) No team scored fewer points than C, which scored 11 points.

(2) Team *A* won two events and came second in the third event.

A. statement 1 alone is sufficient, but statement 2 alone is not sufficient to answer the question

B. statement 2 alone is sufficient, but statement 1 alone is not sufficient to answer the question

C. both statements taken together are sufficient to answer the question, but neither statement alone is sufficient

D. each statement alone is sufficient

E. statements 1 and 2 together are not sufficient, and additional data is needed to answer the question

18. Every pupil in a school was given one ticket for a concert. The school was charged a total of $6000 for these tickets, all of which were of equal valu E. What was the price of one ticket?

(1) If the price of each ticket had been one dollar less, the total cost would have been 1,200 less.

(2) If the price of each ticket had been $2 more, the total bill would have increased by 40%.

A. statement 1 alone is sufficient, but statement 2 alone is not sufficient to answer the question

B. statement 2 alone is sufficient, but statement 1 alone is not sufficient to answer the question

C. both statements taken together are sufficient to answer the question, but neither statement alone is sufficient

D. each statement alone is sufficient

E. statements 1 and 2 together are not sufficient, and additional data is needed to answer the question

19. What is the ratio of male to female officers in the police force in town T?

(1) The number of female officers is 250 less than half the number of male officers.

(2) The number of female officers is 1/7 the number of male officers.

A. statement 1 alone is sufficient, but statement 2 alone is not sufficient to answer the question

B. statement 2 alone is sufficient, but statement 1 alone is not sufficient to answer the question

C. both statements taken together are sufficient to answer the question, but neither statement alone is sufficient

D. each statement alone is sufficient

E. statements 1 and 2 together are not sufficient, and additional data is needed to answer the question

20. What is the value of n?

(1) $3n + 2m = 18$

(2) $n - m = 2n - (4 + m)$

A. statement 1 alone is sufficient, but statement 2 alone is not sufficient to answer the question

B. statement 2 alone is sufficient, but statement 1 alone is not sufficient to answer the question

C. both statements taken together are sufficient to answer the question, but neither statement alone is sufficient

D. each statement alone is sufficient

E. statements 1 and 2 together are not sufficient, and additional data is needed to answer the question

21. How long did it take Henry to drive to work last Wednesday? (He did not stop on the way).

(1) If he had driven twice as fast he would have taken 35 minutes.

(2) His average speed was 30 miles per hour.

A. statement 1 alone is sufficient, but statement 2 alone is not sufficient to answer the question

B. statement 2 alone is sufficient, but statement 1 alone is not sufficient to answer the question

C. both statements taken together are sufficient to answer the question, but neither statement alone is sufficient

D. each statement alone is sufficient

E. statements 1 and 2 together are not sufficient, and additional data is needed to answer the question

22. What is the slope of line l which passes through the origin of a rectangular coordinate system?

(1) The line does not intersect with the line $y = x + 2$

(2) The line passes through the point (3, 3)

A. statement 1 alone is sufficient, but statement 2 alone is not sufficient to answer the question

B. statement 2 alone is sufficient, but statement 1 alone is not sufficient to answer the question

C. both statements taken together are sufficient to answer the question, but neither statement alone is sufficient

D. each statement alone is sufficient

E. statements 1 and 2 together are not sufficient, and additional data is needed to answer the question

IV. Numerical Reasoning (5 questions)

Each question is a sequence of numbers with one or two numbers missing. You have to figure out the logical order of the sequence to find out the missing number(s).

23. 17, 34, 68, 136, ( )

A.168	B.234
C.272	D.257

24. 7, 12, 19, 28, ( )

A.31	B.33
C.35	D.39

25. 11, 12, 23, 35, ( )

A.52	B.58
C.76	D.75

26. -1/2, 1/4, -1/8, 1/16, -1/32, ( )

A.-1/64	B.1/64
C.1/128	D.1/128

27. 9, 8.8, 7.8, 6.9, ( )

A.5.9	B.6
C.6.1	D.6.2

V. Interpretation of Tables and Graphs (8 questions)

This is a test on reading and interpretation of data presented in tables and graphs.

Graph 1 (Question 28 to 30)

City	Average hours worked per year per person	Average Days holiday taken	Size of work-force
London	1,785	20	7,500,000
Madrid	1,758	22	3,000,000
Rome	1,747	21	57,500,000
Dublin	1,727	21	500,000
Stockholm	1,726	25	1,500,000
Athens	1,704	24	32,000,000
Amsterdam	1,687	25	750,000
Oslo	1,627	24	500,000

28. What is the average hours worked per year per person for Oslo and Athens combined?

A. 3,331　　B. 1,703

C. 1,782　　D. 1,679

29. A working day in Dublin consists of eight hours. If the average number of days holiday taken were to be reduced by two, what would be the new average hours worked in Dublin per person per year?

A. 1,743　　B. 1,726

C. 1,738　　D. 1,711

30. If the average hours worked per year per person in Dublin were the same as that of London, how many more hours would be worked in Dublin in a year?

A. 29 million　　B. 26 million

C. 31 million　　D. 18 million

Graph 2 (Question 31 to 33)

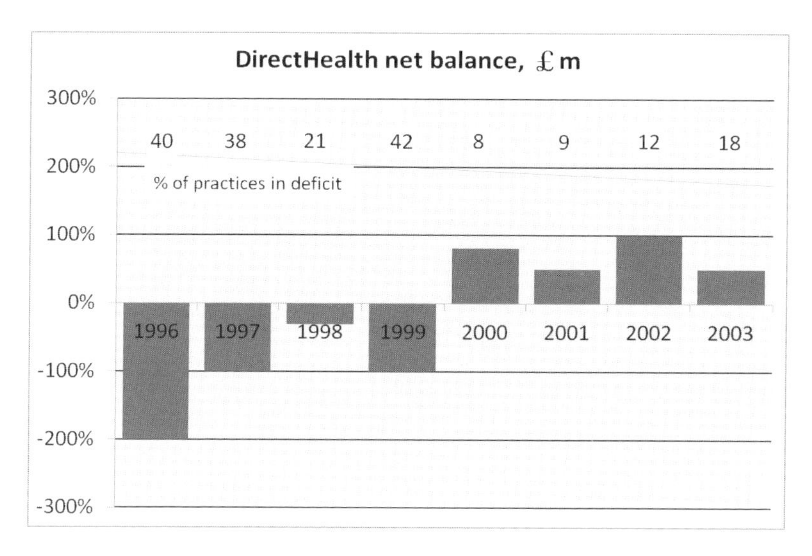

31. If in 2002 there were 50 DirectHealth practices, how many were not in ?

    A. 44 B. 21 C. 10 D. Cannot Say

32. In 1997, 21 practices were all in profit by an average of £2m each. What was the average deficit run by each of the remaining 39 practices?

    A. £3.6m B. £1.2m C. £24m D. £180m

33. How much more money would DirectHealth have had to have made in 1999 to equal the net balance of 2001?

    A. £25m B. £150m C. £1m D. £52m

Graph 3 (Question 34 to 35)

The pie chart below represents the monthly distribution of family X's income.

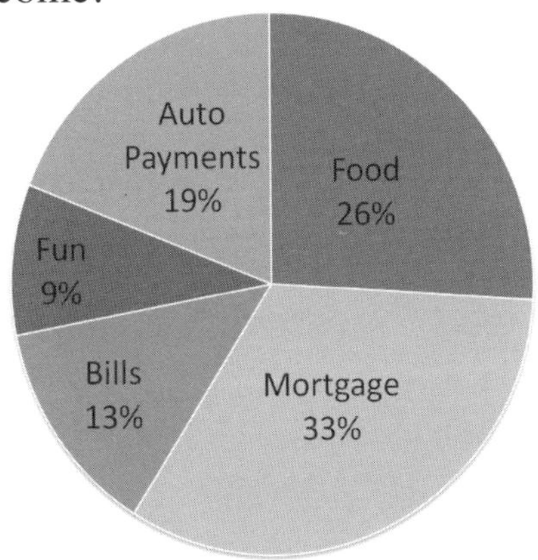

Monthly Household Income = $2,400

34. According to the graph, how much is family X's mortgage in dollars?

    A. 216 B. 312 C. 456 D. 624 E. 792

35. If family X completed all its car payments and transferred all the auto money into the "Fun" budget, then family X would most nearly spend the same amount on fun as they spend on

    A. food
    B. mortgage
    C. bills
    D. food and bills
    E. food and mortgage

ANSWERS

I. 演繹推理 (8題)

1. 答案：D。題幹中的論點是“檢驗該司機走直線的能力是檢驗其是否適於駕車的更可靠的指標”，A項顯然與詰一論點相悖，而B項顯然是違背常識的，C項的説泓也明顯與論點相左，所以只能選D項。血液酒精含量高有可能仍能駕車，所以這種的可參考性低，正好説明了論點。

2. 答案：C。由於有三種牌共20張，如果其中有兩種總數超過了19，也就是達到20張，那麼另外一種牌就不存在了，可見〔3〕的説法正確；假設兩種牌的張數分別是：6、6、8，就推翻了〔1〕的假設，所以〔1〕不正確，因此，只能選C。

3. 答案：D。小李的主要觀點是：由於廣告廠商通過提過提高商品價格，將高額的廣告費用轉嫁到觀眾頭上，這樣奧運會的觀眾沒有從拉的廣告中得到經濟利益。要削弱這種説法，就要找出與之相悖的選項。A項顯然不能；B項雖然是事實，但卻不能推翻小李的説法；C恰好証明了小李的觀點。由於廣告支出是常量，所以廣告商也不會提高商品價格，消費者自然也就能從中獲益，所以，D項正確。

4. 答案：A。用戶的不滿不會對節約用水產生直接性的影響，（3）應該排除，也就説，B、C、D三項都不能選。

5. 答案：B。題幹中指出，高等院校在招收新老師時，除了具有博士學位的，還有少數優勢的碩士畢業生，A項論述錯誤；C項在題幹中沒有反映，題幹中説明；事實上，大多數它的教授並沒有博士學位，所以D項選法錯誤。

6. 答案：D。動物的價值並不能簡單的判定，所以A、B的説法都是不正確的；同時題中只強調保護瀕臨滅絕動物需要高額費用，並沒有説支付這些費用是不正確的，故C項錯誤。

7. 答案：B。題中的75個嬰兒在長大後偏側性發生變化的只有5個，説明大部分人的偏側性並不會隨年齡的增長而變化，B不能從正文中推導出來。

8. 答案：D。甲的觀點是閱讀導致近視眼，乙的觀點是因為有近視眼看不清遠處，所以選擇近距離可以看清的活動——閱讀。所以答案是D。

II. Verbal Reasoning (English) (6 questions)

9. B (False): Buddhism was adopted by the court nobility at the urging of the emperor.

10. B (False): The introduction of Buddhism to Japan led to great political unrest and social confusion.

11. C (Can’t tell): Buddhism replaced the Shinto religion which had previously been followed in Japan.

12. A (True): Japanese men do not share household chores and childcare with their wives.

13. B (False): A quarter of all part-time workers in Japan are female.

14. A (True): Part-time workers hold a low status in Japanese companies.

III. Data Sufficiency Test (8 questions)

15. A 16. D 17. C 18. D

19. B 20. B 21. A 22. D

IV. Numerical Reasoning (5 questions)

23. C 24. D 25. B 26. B 27. C

V. Interpretation of Tables and Graphs (8 questions)

28. B: You need to first find the total number of hours worked by the workforce in Oslo (1,627x500,000) and the same for Athens (1,704x32,000,000). Then divide this by the combined population (500,000+32,000,000).

29. A: With two days less holiday, each person would work 16 hours more per year, so hours worked per person per year would go up by 16.

30. A: Calculate (size of workforce) x (difference in hours per year per person). (500,000) x (1782-1727).

31. A: The figure says that 12% of practices were in deficit. Therefore 88% of 50 were not in deficit.

32. A: 21 practices each in profit by £2m means the remaining 39 practices need between them to be in a deficit of £142m to cause the net balance to be -£100 as shown.

33. B: Interpreting the graph, it looks as though the net balance in 2001 was £50m. The difference between £50m and -£100m is £150m.

34. E

35. A

能力傾向測試
模擬測驗（四）

限時四十五分鐘

I. 演繹推理(8題)

請根據以下短文的內容，選出一個或一組推論。請假定短文的內容都是正確的。

1. 對於穿鞋來說，正合腳的鞋子比大一些的鞋子好。不過，在寒冷的天氣，尺寸稍大點的毛衣與一件正合身的毛衣差別並不大。這意味着：

   A. 不合腳的鞋不能在冷天穿

   B. 毛衣的大小只不過是式樣的問題，與其功能無關

   C. 不合身的衣物有時仍然有使用價值

   D. 在買禮物時，尺寸不如用途那樣重要

2. 有文章指出，嬰兒對音樂或歌聲的反應要比話語更強烈。對於6個月大的嬰兒，母親的歌聲是最容易讓其入眠的。同時，音樂旋律能夠反映人體自身的韻律，比如，心跳和呼吸的節律。根據這段話，以下做法合理的是：

   A. 用音樂為嬰兒治療生理性疾病

   B. 用音樂進行胎教，鍛煉胎兒的四肢

   C. 用音樂調節嬰兒的情緒和反應

   D. 用兒童喜歡的音樂旋律推測其身體狀況

3. 自然界的水因與大氣、土壤、岩石等接觸，所以含有多種“雜質”，如鈣，鎂、鉀、鈉、鐵、氟等。現代人趨向於飲用越來越純淨的水，如蒸餾水、純水、太空水等。殊不知，長期飲用這種超純淨的水，會不利於健康。

下列選項最能支持上述論斷的是：

A. 人們對飲食衛生越注重，人體的健康就會越脆弱

B. 只有未經處理的自然界的水，才符合人體健康的需要

C. 超純淨水之所以大受歡迎，是因為它更加衛生、口感更好

D. 自然界水中的所謂”雜質”，可能是人體必需微量元素的重要來源

4. 一般來說，科學家在進行科學研究時，容易被與其目標一致的其他科學家所接受，作為他們的同事。而當某位科學家作為向大眾解釋科學的人獲得聲譽時，大多數其他科學家會認為他不能再被視為一位真正的同事了。

以上論斷所基於的假設：

A. 嚴肅的科學研究不是一項個人的活動，而是要依賴一群同事的積極協作

B. 從事研究的科學家們不把他們所嫉妒的有名的科學家們視為同事

C. 一位科學家可以在沒有完成任何重要研究的情況下成為一位知名人士

D. 從事研究的科學家們認為那些科學名人沒有動力去從事重要的新研究

5. 對那些很少刷牙的人來說，罹患口腔癌的危險性更高。為了能在早期發現這些人的口腔癌，某市衛生部門向該市居民散發了小冊子，上面描述了如何進行每周口腔的自我檢查以發現口腔內的腫瘤。

下面哪個選項如果正確，最能質疑上述做法的效果？

A. 口腔癌在成年人中比在兒童中更為普遍

B. 這份小冊子被散發到該市的所有居民

C. 很少刷牙的人不大可能每周對他們的口腔進行檢查

D. 許多牙病症狀是不能在每周自我檢查中被發現的

6. 有三位男生張強、趙林、王剛和三位女生李華、秦珊、劉玉暑假外出旅游，可能去上海、杭州、青島和大連。條件是：（1）每人只能去一個地方；（2）凡是男生去的地方就必須有女生去；（3）凡是有女生去的地方就必須有男生去；（4）李華去上海或者杭州，趙林去大連。

如果上述斷定都為真，則去杭州的人中不可能同時包含：

A. 張強和王剛

B. 王剛和劉玉

C. 秦珊和劉玉

D. 張強和秦珊

7. 在一種插花藝術中，對色彩有如下要求：（1）或者使用橙黃或者使用墨綠；（2）如果使用橙黃，則不能使用天藍；（3）只有使用天藍，才使用鐵青；（4）綠和鐵青只使用一種。

由此可見在該種插花藝術中：

A. 不使用墨綠，使用天藍

B. 不使用橙黃，使用鐵青

C. 不使用鐵青，使用墨綠

D. 不使用天藍，使用橙黃

8. 如果生產下降或浪費嚴重，那麼將造成物資匱乏。如果物資匱乏，那麼或者物價暴漲，或者人民生活貧困。如果人民生活貧困，政府將失去民心。事實上物價沒有暴漲，而且政府贏得了民心。

由此可見：

A. 生產下降但是沒有浪費嚴重

B. 生產沒有下降但是浪費嚴重

C. 生產下降並且浪費嚴重

D. 生產沒有下降並且沒有浪費嚴重

II. Verbal Reasoning (English) (6 questions)

In this test, each passage is followed by three statements (the questions). You have to assume what is stated in the passage is true and decide whether the statements are either:

(A) True: The statement is already made or implied in the passage, or follows logically from the passage.

(B) False: The statement contradicts what is said, implied by, or follows logically from the passage.

(C) Can't tell: There is insufficient information in the passage to establish whether the statement is true or false.

Passage 1 (Question 9 to 11)

So much of the literature of the western world, including a large part of its greatest literature, was either written for actual speaking or in a mode of speech, that we are likely to deform it if we apply our comparatively recent norm of writing for silent reading. It is only that so much of this work is drama or oratory (the latter including the modern forms of sermons, lectures and addresses which as late as the nineteenth century play a most important part). It is also that through classical and mediaeval times, and in many cases beyond these, most reading was either aloud or silently articulated as if speaking: a habit we now recognize mainly in the slow. Most classical histories were indeed quite close to oratory and public speech, rather than silent reading of an artifact, was the central condition of linguistic composition.

9. Until the nineteenth century, most people could only read with difficulty.

10. In ancient times, literature was intended to be read aloud.

11. Classical histories were passed on orally and never written down.

Passage 2 (Question 12 to 14)

Millions of lives around the world could be saved, and the quality of life of hundreds of millions markedly improved - very inexpensively - by eradicating three vitamin and mineral deficiencies in people's diets. The three vitamins and minerals are vitamin A, iodine and iron - so-called micronutrients. More than 2 billion people are at risk from micronutrient deficiencies and more than 1 billion people are actually ill or disabled by them, causing mental retardation, learning disabilities, low work capacity and blindness. It costs little to correct these deficiencies through fortification of food and water supplies. In a country of 50 million people, this would cost about $25 million a year. That $25 million would yield a forty fold return on investment.

12. Most illnesses in developing countries are caused by vitamin and mineral deficiencies.

13. Micronutrients provide inadequate nourishment to maintain a healthy life.

14. Vitamin A, iodine and iron are the only micronutrients that people need in their diet.

III. Data Sufficiency Test (8 questions)

In this test, you are required to choose a combination of clues to solve a problem.

15. If x, y and z are different integers, is x divisible by 11?

    (1) xyz is divisible by 22 and 33

    (2) yz is divisible by 72

    A. statement 1 alone is sufficient, but statement 2 alone is not sufficient to answer the question
    B. statement 2 alone is sufficient, but statement 1 alone is not sufficient to answer the question
    C. both statements taken together are sufficient to answer the question, but neither statement alone is sufficient
    D. each statement alone is sufficient
    E. statements 1 and 2 together are not sufficient, and additional data is needed to answer the question

16. A die is rolled randomly on to a circular board with a triangle inscribed in the circle. (All three vertices of the triangle are on the circumference of the circle.) What is the probability that the die comes to rest outside the triangular region?

(1) The hypotenuse of the triangle is a diameter of the circle.

(2) The radius of the circle is 2 units, and the area of the triangle is 4 square units.

A. statement 1 alone is sufficient, but statement 2 alone is not sufficient to answer the question

B. statement 2 alone is sufficient, but statement 1 alone is not sufficient to answer the question

C. both statements taken together are sufficient to answer the question, but neither statement alone is sufficient

D. each statement alone is sufficient

E. statements 1 and 2 together are not sufficient, and additional data is needed to answer the question

17. A university has 2,000 faculty members all of whom have a masters degree, and some of whom also have a doctorate. 35 percent of the faculty members are female. What fraction of the faculty members are female doctorate holders?

(1) 20 percent of the male faculty members have a doctorate.

(2) A total of 1140 faculty members have only a masters degree.

A. statement 1 alone is sufficient, but statement 2 alone is not sufficient to answer the question

B. statement 2 alone is sufficient, but statement 1 alone is not sufficient to answer the question

C. both statements taken together are sufficient to answer the question, but neither statement alone is sufficient

D. each statement alone is sufficient

E. statements 1 and 2 together are not sufficient, and additional data is needed to answer the question

18. The retail price of a certain refrigerator was Z dollars. Nina was given a further 10 percent discount on the already discounted sale price (Y) of this refrigerator. Given that the price Nina paid was X dollars, what was the dollar value of the extra 10 percent discount that she obtained?

    (1) The sale price, Y, was 90 percent of the normal retail price, Z.

    (2) If the sale price, Y, had been 20 dollars more, then X would have been \$14 less than this new value of Y.

A. statement 1 alone is sufficient, but statement 2 alone is not sufficient to answer the question

B. statement 2 alone is sufficient, but statement 1 alone is not sufficient to answer the question

C. both statements taken together are sufficient to answer the question, but neither statement alone is sufficient

D. each statement alone is sufficient

E. statements 1 and 2 together are not sufficient, and additional data is needed to answer the question

19. In triangle ABC all the sides have integer lengths. What is the length of side AC?

(1) $AB = 3$ and $BC = 4$

(2) One of the angles of ABC is a right angle

A. statement 1 alone is sufficient, but statement 2 alone is not sufficient to answer the question

B. statement 2 alone is sufficient, but statement 1 alone is not sufficient to answer the question

C. both statements taken together are sufficient to answer the question, but neither statement alone is sufficient

D. each statement alone is sufficient

E. statements 1 and 2 together are not sufficient, and additional data is needed to answer the question

20. Is $x < 0$?

(1) $7x > 8x$

(2) $-3(x) > 0$

A. statement 1 alone is sufficient, but statement 2 alone is not sufficient to answer the question

B. statement 2 alone is sufficient, but statement 1 alone is not sufficient to answer the question

C. both statements taken together are sufficient to answer the question, but neither statement alone is sufficient

D. each statement alone is sufficient

E. statements 1 and 2 together are not sufficient, and additional data is needed to answer the question

21. In which year was Heidi born?

(1) Heidi's daughter was born in 1960 when Heidi was 28 years old.

(2) Heidi's birthday and her daughter's birthday are exactly six months apart.

A. statement 1 alone is sufficient, but statement 2 alone is not sufficient to answer the question

B. statement 2 alone is sufficient, but statement 1 alone is not sufficient to answer the question

C. both statements taken together are sufficient to answer the question, but neither statement alone is sufficient

D. each statement alone is sufficient

E. statements 1 and 2 together are not sufficient, and additional data is needed to answer the question

22. Is $xy < 15$?

(1) $0.5 < x < 1$, and $y^2 = 144$

(2) $x < 3$, $y < 5$

A. statement 1 alone is sufficient, but statement 2 alone is not sufficient to answer the question

B. statement 2 alone is sufficient, but statement 1 alone is not sufficient to answer the question

C. both statements taken together are sufficient to answer the question, but neither statement alone is sufficient

D. each statement alone is sufficient

E. statements 1 and 2 together are not sufficient, and additional data is needed to answer the question

IV. Numerical Reasoning (5 questions)

Each question is a sequence of numbers with one or two numbers missing. You have to figure out the logical order of the sequence to find out the missing number(s).

23. 2/1, 4/3, 9/4, 16/5, 25/6, ( )

A. 31/7 B. 33/7
C. 36/7 D. 29/7

24. 2.1, 4.2, 7.3, ( ), 18.8

A. 10.6 B. 11.2
C. 13.5 D. 11.5

25. 1/6, 1, 5, 20, 60, ( )

A. 180 B. 150
C. 120 D. 90

26. 11, 23, 35, 47, 59, ( )

A. 61 B. 63
C. 71 D. 73

27. 5/2, 9/4, 13/6, ( )

A.15/8 B.2
C.17/8 D.4/9

V. Interpretation of Tables and Graphs (8 questions)

This is a test on reading and interpretation of data presented in tables and graphs.

Graph 1 (Question 28 to 30)

A student walks to the bus stop to catch a bus to the university. He then walks from the bus stop at the university to the students union arriving there at 8.35am.

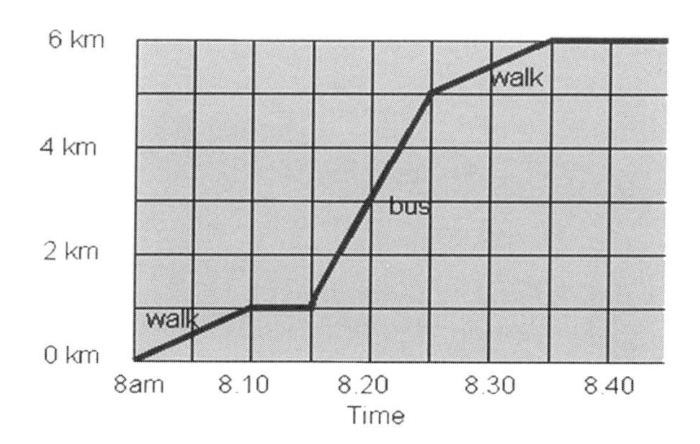

28. How far does the student walk in total?

    A. 1km B. 2km C. 3km D. 4km E. 5km

29. How far is he from the university students' union at 8.20 am?

    A. 1km B. 2km C. 3km D. 4km E. 5km

30. What is the average speed of the bus?

A. 14kmph B. 24kmph C. 32kmph

D. 40kmph E. 48kmph

First Destinations of Students from the University of Poppleton			
	2002	2003	2004
Total no. of grads	1,700	1,600	1,500
% in employment	40	38	37
% in further study	33	36	39
% unemployment	4	6	8
% other	23	20	17

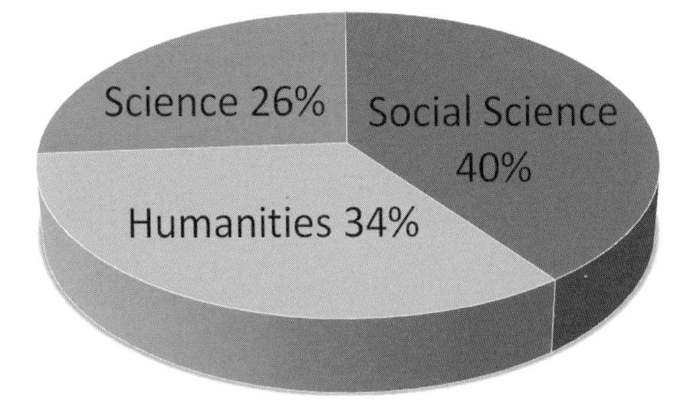

Graph 2 (Question 31 to 33)

31. What was the largest number of students in any year that went on to further study?

A. 561 B. 576 C. 585 D. 592 E. Can't Say

32. What was the decrease in the number of graduates in employment between 2002 and 2004?

A.125 B. 135 C. 140 D. 180 E.Can't Say

33. In 2004 how many social science students were in employment after graduating?

A. 260 B. 272 C. 284 D. 290 E. Can't Say

Graph 3 (Question 34 to 35)

The graph below gives the number of computers sold each month (in thousands) by three different computer manufacturers Manufacturer 1 (M1), Manufacturer 2 (M2) and Manufacturer 3 (M3).

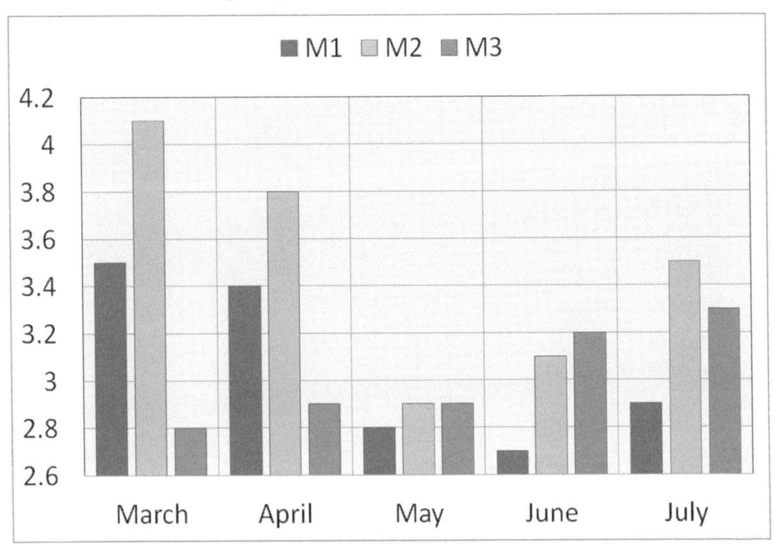

34. Which month showed the largest total decrease in PC sales over the previous month?

A. March B. April C. May D. June E. July

35. What percentage of Manufacturer 2's sales were made in April (to the nearest percent)?

A. 16 B. 22 C. 27 D.33 E. 38

ANSWERS

I. 演繹推理 (8題)

1. 答案： C。只有C是可以從陳述中直接推出的，應選C。

2. 答案： C。演繹推理題的關鍵在於選擇古要能從題幹的陳述中直接推出。本題中，A、B、D都不能從陳述中直接推出，故選C

3. 答案： D。題幹中講的是長期飲用缺少必要的礦物質的純淨水會不利於健行。故D為最符合的答案。

4. 答案： D。本題要根據題目中的信息講行合理的推斷，顯然，D項最符合題意。

5. 答案： D。本題要求的是最佳答案，比較C、D，D是最佳答案。

6. 答案： C。根據題意，根據題意，由於趙林去大連，所以根據(2)可知三舍女生中必須有人去大連，而根據(4)李華去上海或杭州，則可推知另外兩個女生秦珊和劉玉必須有人去大連，所以秦珊和劉玉不能同時去杭州。

7. 答案： D。根據題意，假設使用墨綠，則根據(4)，可推知不使用鐵青，又根據(3)可推知不使用鐵青，則也不使用天藍；進而又根據(2)可推知不使用天藍，則使用橙黃，這與(1)矛盾，故不使用墨綠，則使用橙黃，故選D。

8. 答案： D。根據題意，因為政府贏得了民心，則人民生活不貧困，並且物價沒有暴漲，所以物資不匱乏，所以生產沒有下降並且沒有浪費嚴重，故選D。

II. Verbal Reasoning (English) (6 questions)

9. C (Can't Say): Until the nineteenth century, most people could only read with difficulty.

10. A (True): In ancient times, literature was intended to be read aloud.

11. B (False): Classical histories were passed on orally and never written down.

12. C (Can't Say): Most illnesses in developing countries are caused by vitamin and mineral deficiencies.

13. B (False): Micronutrients provide inadequate nourishment to maintain a healthy life.

14. C (Can't Say): Vitamin A, iodine and iron are the only micronutrients that people need in their diet.

III. Data Sufficiency Test (8 questions)

15. E	16. B	17. C	18. C
19. D	20. E	21. E	22. A

IV. Numerical Reasoning (5 questions)

23. C	24. D	25. C	26. C	27. C

V. Interpretation of Tables and Graphs (8 questions)

28. B	29. C	30. B	31. C
32. A	33. E	34. C	35. B

PART
FOUR

考生急症室——

1）每隔多久考CRE一次？

CRE一年考兩次，分別在6月和10月考試。

2）什麼人符合申請資格？

- 持有大學學位（不包括副學士學位）；或
- 現正就讀學士學位課程最後一年；或
- 持有符合申請學位或專業程度公務員職位所需的專業資格。

3）若然在香港中學文憑考試英國語文科及／或中國語文科取得5級或以上成績，是否需要報考綜合招聘考試英文運用及／或中文運用試卷？

香港中學文憑考試英國語文科5級或以上成績會獲接納為等同綜合招聘考試英文運用試卷的二級成績。香港中學文憑考試中國語文科5級或以上成績會獲接納為等同綜合招聘考試中文運用試卷的二級成績。持有上述成績者不須考試。

4）若然在香港高級程度會考英語運用科（或General Certificate of Education A Level (GCE A Level) English Language科）及／或中國語文及文化科取得及格成績，可否獲豁免參加綜合招聘考試？

香港高級程度會考英語運用科或GCE A Level English Language科C級或以上成績會獲接納為等同綜合招聘考試英文運用試卷的二級成績；香港高級程度會考中國語文及文化、中國語言文學或中國語文科C級或以上成績會獲接納為等同綜合招聘考試中文運用試卷的二級成績。如果持有上述成績者不須考試。

香港高級程度會考英語運用科或GCE A Level English Language科D級成績會獲接納為等同綜合招聘考試英文運用試卷的一級成績；香港高級程度會考中國語文及文化、中國語言文學或中國語文科D級成績會獲接納為等同綜合招聘考試中文運用試卷的一級成績。如果持有上述成績，可因應有意投考的公務員職位的要求，決定是否需要應考綜合招聘考試英文運用及/ 或中文運用試卷。

5）「綜合招聘考試」(CRE)跟「聯合招聘考試」(JRE)有何分別？

在CRE中英文運用考試中取得「二級」成績後，可投考JRE，考試為AO、EO及勞工事務主任、貿易主任四職系的招聘而設。

6） CRE成績何時公佈？

考試邀請信會於考前12天以電郵通知，成績會在試後1個月內郵寄到考生地址。

7） 報考CRE的費用是多少？

不設收費。

8） 若然在綜合招聘考試的英文運用及中文運用試卷取得二級或一級成績，並在能力傾向測試中取得及格成績，是否已符合資格申請公務員職位？可以在何時及怎樣申請這些職位？

個別進行招聘的部門/ 職系會於招聘廣告中列明有關職位所需的綜合招聘考試成績。由於綜合招聘考試與公務員職位的招聘程序是分開進行的，應留意在各報章及公務員事務局網頁刊登的公務員職位招聘廣告，然後直接向進行招聘的部門/ 職系提交職位申請。進行招聘的部門/ 職系會核實你的學歷及/ 或專業資格，並可能在綜合招聘考試外，另設其他考試/ 面試。

9） 可否使用CRE的成績來申請政府以外的工作？

CRE招聘考試是為招聘學位或專業程度公務員職位而設的基本測試，而非一項學歷資格。

10） 如遺失了CRE考試／基本法測試的成績通知書，可否申請補領？

可以書面（地址：香港添馬添美道2號政府總部西翼7樓718室）或電郵形式（電郵地址：csbcseu@csb.gov.hk）向公務員考試組提出申請。

看得喜 放不低

創出喜閱新思維

書名　投考公務員 能力傾向測試精讀王　修訂第二版
ISBN　978-988-78873-6-2
定價　HK$88 / NT$280
出版日期　2018年8月
作者　Man Sir & Mark Sir
責任編輯　投考紀律部隊系列編輯部
版面設計　梁文俊
出版　文化會社有限公司
電郵　editor@culturecross.com
網址　www.culturecross.com
發行　香港聯合書刊物流有限公司
地址：香港新界大埔汀麗路36號中華商務印刷大廈3樓
電話：（852）2150 2100
傳真：（852）2407 3062
台灣總經銷：貿騰發賣股份有限公司
電話：（02）822 75988